Shipwreck

Dr Sam Willis is one of the world's leading authorities on maritime and naval history. He is a visiting Fellow at the University of Plymouth and a Fellow of the Royal Historical Society. He has consulted widely on maritime affairs for clients including the BBC, Channel 4 and Christie's. He is the author of several books on naval and maritime history, including the 'Hearts of Oak' trilogy and the 'Fighting Ships' series.

www.sam-willis.com

Shipwreck

A History of Disasters at Sea

Sam Willis

New York • London

Quercus
New York • London

First published in the United States by Quercus in 2015

ISBN 978-1-62365-479-5

Library of Congress Control Number: 2015938655

Distributed in the United States and Canada by
Hachette Book Group
1290 Avenue of the Americas
New York, NY 10104

Manufactured in the United States

2 4 6 8 10 9 7 5 3 1

www.quercus.com

Contents

Introduction

Methoughts I saw a thousand fearful wrecks; / Ten thousand men that fishes gnaw'd upon; Wedges of gold, great anchors, heaps of pearl, / Inestimable stones, unvalu'd jewels, All scatter'd in the bottom of the sea. / Some lay in dead men's skulls, and in the holes / Where eyes did once inhabit, there were crept – / As 'twere in scorn of eyes – reflecting gems, / That woo'd the slimy bottom of the deep, / And mock'd the dead bones that lay scatter'd by.

WILLIAM SHAKESPEARE – CLARENCE'S DREAM,
RICHARD III ACT I: SCENE IV

In December 1790 an American ship was wrecked on the sandbanks that lie between the mouth of the River Mersey and the Irish Channel. The crew clung to the remains of their vessel, but there was only enough room for two of the eleven survivors to perch on the wreckage that was clear of the water. The two men who chose the exalted positions were the ship's master and the only paying passenger. The rest of the crew did as best they could in the semi-submerged wreck, and the single black sailor aboard was forced to wait for rescue with only his head exposed. Fortunately rescue soon came, as their distress signal had been seen from nearby Hillberry Island. By the time that they were rescued, however, 23 hours after their ship had sunk, only nine remained alive. One account tells how the master and the paying passenger, both strong and healthy men and the only two men not submerged, had managed to grab hold of a keg of cherry brandy as the ship sank. The keg still contained the brandy-infused cherries which the two men promptly devoured. Both dropped dead within an hour or two.

The Enduring Fascination of Shipwrecks

That this story has survived is indicative of our longstanding fascination with shipwrecks. For contemporary sailors, the death of the two men was easily explained: their intoxication clearly hastened their death. Spirits, many sailors believed, fortified the body against the effects of heat, moisture and fatigue, but was always hurtful in severe and continued cold. Sailors also believed that wet clothes helped stave off thirst. On one hand, then, the story confirmed the sailors' traditional beliefs and was retained and retold as proof. On the other, it served a much wider purpose as an allegorical story illustrating the evils of drink: it is exactly the kind of shipwreck story that so pleased a mid-Victorian audience high on the benefits of temperance. There are also echoes in the narrative of darker themes of racial prejudice and the privileges of wealth, which would further have encouraged its re-telling. Still further, the original story fired a young doctor named James Curry with enthusiasm, for he fervently believed that the cherries had no bearing on either of the men's deaths. He determined to prove scientifically that shipwrecked sailors in wet weather and cold climates were better advised to stay submerged in salt water than to risk exposure in clothes drenched by fresh water. He published his detailed experiments in which he made a number of unfortunate young men endure freezing temperatures in various states of dress and for lengthy periods of time in the *Transactions of the Royal Society of London* in 1792. It makes extraordinary reading.

This single and relatively uneventful shipwreck therefore appealed greatly to a wide range of people for a number of unrelated reasons, and therein lies our timeless fascination with shipwreck, a fascination that has readily been met by authors, artists, historians and archaeologists for generations. It has been remarkably difficult, therefore, to select the shipwrecks for this book, quite simply because there are so many of significance to chose from. Nevertheless, I believed that there were some important requirements that had to be met. Iconic shipwrecks such as the *Mary Rose* and *Titanic* needed inclusion, but at the same time it was important to include less well known but equally significant wrecks, such as that of the *Wilhelm Gustloff*. Largely unknown outside the confines of maritime history, the *Wilhelm Gustloff* is by some margin the worst disaster in maritime history; almost six times more people died on the *Gustloff* than on the *Titanic*.

It was also important to provide both an even geographical and chronological coverage for periods that are traditionally lopsided, such as the Dark Ages. While many will have heard of the discovery of Viking ships, very few indeed know anything of the explosion of seafaring that was underway at the same time in Southeast Asia, and only recently is maritime archaeology beginning to answer so many of our questions for that period. Otherwise, I have been careful to include merchantmen, pirates and warships; ships and submarines; ships that were sunk through human folly, and ships that were sunk through warfare; individual wrecks and mass wrecks such as those at Scapa Flow, where 72 ships of the German High Seas Fleet were scuttled in the twilight of the First World War. In every case, however, the wrecks have been included because they say something important about global, rather than purely maritime, history. The Skuldelev ships, for example, are more than just a collection of Viking ships; their discovery revolutionized the way we understand the development of northern Europe itself under the Vikings. Similarly, the magnificent wreck of the *Vasa* is a powerful reminder of the importance of Baltic seapower in the development of early modern Europe, all too often overlooked by history. The *Indianapolis*, wrecked at the very end of the Second World War, is usually remembered for the hundreds of men taken by sharks as they drifted across the Pacific, but her story is intricately linked with the explosion of the first-ever atomic bomb. Finally, the sinking of the Russian submarine *Kursk* in 2000 says as much about the decline of post-Soviet Russia and international relations after the Cold War as it does about the deaths of her brave crew.

As with all shipwrecks, therefore, it is possible and rewarding to appreciate them on many different levels. Most obviously, tales of shipwreck are attractive as straightforward descriptions of human tragedy, in which heroism, sacrifice and villainy all bubble to the surface in conditions of extreme physical and emotional suffering. Humans are unable to survive for long in open water. They cannot drink seawater and quickly begin to suffer from hypothermia and cramp. Struggling to stay afloat, they make easy targets for sea creatures. Accounts litter the history of seafaring of attacks from fish, sharks, jellyfish, octopus and seals; even tiny crabs work their way into open wounds and salt-water ulcers. Often forgotten, the danger from seabirds is also very real. A prolonged attack will force the swimmer to tire quickly and raising his arms in defence of his head and eyes will cause him to sink. Extended exposure in the sun without sustenance can lead to hallucinations. Feelings of panic, despair, forgetfulness and humiliation are all common to sufferers of shipwreck who have been reduced to their most basic instincts for survival. Almost all suffer from shock in some form.

Large ships create a dangerous suction when they submerge and they can pull survivors down with them, even forcing them back into the hull as water pours into every vacant chamber. Those left on the surface are then vulnerable to items resurfacing at speed, often in air bubbles, as they free themselves from the stricken ship. Cut feet and broken ankles are common injuries in this scenario. Survivors who fall to leeward of the stricken ship are often crushed by the hull as, driven by the wind or current, the ship can move more quickly to leeward than it is possible to swim. Those who make it to shore are still vulnerable if there is breaking surf and strong tidal surges near the shore. These facts are as true now as they were for the very first seafarers.

However, regardless of the form a shipwreck story takes, whether it concentrates on human suffering or not, the need for authenticity in its depiction remains central to its success. In 1762 William Falconer, himself the survivor of the wreck of the merchantman *Britannia*, published his poem 'The Shipwreck':

'Again she plunges! Hark! A second shock
Bilges the splitting vessel on the rock
Down on the vale of death, with dismal cries
The fated victims cast their shuddering eyes
In wild despair; while yet another stroke
With strong convulsions rends the solid oak.'

It is not the work of a literary genius, but nevertheless, it was riddled throughout with maritime terms and bruising descriptions which lent an authenticity to the poem. Moreover, Falconer actually died only seven years later in another shipwreck, when the *Aurora* was lost with all hands, and his death only raised his profile further; this was a man who knew his subject: how much closer could one get to a shipwreck than reading the words of a man who had been wrecked twice, and had even died in a wreck? By 1830, 24 editions of his poem had been published.

Indeed, public desire to get as close as possible to the experience of shipwreck from the safety of a comfortable chair explains in large part the phenomenal success of James Cameron's 1997 film *Titanic*, which grossed more at the box office than any other film in history. Its success owed much to the realism of the shipwreck scenes, filmed in a specially designed tank, 27 metres (90 ft) deep, 244 metres (800 ft) wide, and containing 17 million gallons of water, into which the film-makers sank a 90 percent scale model of the *Titanic*, built from the original Harland & Wolff blueprints.

Such commercial profiteering from shipwreck has a long and illustrious heritage which is little known. As early as 1781, the artist Philippe de Loutherbourg (1740–1812), responsible for some of the most sensationalist depictions of life at sea ever painted, opened his fantastical *Eidophusikon*, a sort of mechanized theatre, which through the use of magic lantern slides, silk filters, clockwork automata and other devices, achieved an unprecedented realism in its depiction of shipwreck and natural phenomena. One of its greatest successes was a rendition of the 1786 sinking of the *Halsewell*. Much like Cameron's film *Titanic*, more than two centuries later, Loutherbourg's *Halsewell* was promoted and celebrated as a more accurate and more realistic depiction of shipwreck than had ever before been created. There was certainly a long and impressive heritage of shipwreck art upon which it drew. Tintoretto's *St Mark Rescuing the Saracen* (1562) is perhaps the first image of a shipwreck where the vast terror of the sea is at all realized and Brueghel's famous *Storm at Sea* (1568–9) is one of the earliest attempts to render the heaving of the deep.

Long after the success of Loutherbourg's *Eidophusikon* had been celebrated, the depiction of shipwreck remained an attractive challenge for generations of artists. Joseph Vernet (1714–89) and J.M.W. Turner (1775–1851) took it to a new level at the turn of the century, and are reputed to have passed though storms on ships and insisted on being tied to the mast. In the words of Vernet, this allowed him '*the better to contemplate and study the imposing majesty of the turmoil of the elements*'. Turner's resulting *The Shipwreck: Fishing Boats Endeavouring to Rescue the Crew* of 1805 is considered to be one of the best depictions of the sea in a storm that had, by then, ever been painted.

Courage and Cowardice *In Extremis*

Shipwrecks can also be read as sophisticated allegories of contemporary issues in which the sinking itself and the death of those aboard are laden with symbolism and significance that extend far beyond the basic facts. One of the most powerful examples comes from the wreck of the *Droits de l'Homme* (The Rights of Man), a French warship that sank in January 1797, in the tumultuous wake of the French Revolution. The sinking of a ship bearing that name was itself significant to those who were critical of the Revolution, but the details of her sinking made the tale even more powerful as the notion of the Rights of Man directly affected the outcome of the wreck: when the largest lifeboat was launched, attempts by the officers to restore order failed hopelessly and over 120 men leapt into the boat, which immediately capsized.

In the late 19th and early 20th centuries, shipwreck came

to the forefront of the national consciousness for similar reasons. In February 1852 the troopship HMS *Birkenhead* foundered on her way to South Africa with over 600 men, women and children aboard. There were insufficient lifeboats and precious little time; the *Birkenhead* was made of iron and she was sinking very quickly. The newspapers recorded how the British soldiers stood calmly in rank as the ship sank under them, while the women and children were saved. It came to be one of the most enduring stories of Victorian military heroism, and it is even believed that Kaiser Wilhelm II had the story read to his troops during the First World War as an example of manly courage. For the British, however, it demonstrated much more than mere manly courage; it demonstrated *British* courage. An excerpt from a typical poem from the period runs:

> *'I've an English ship and an English crew who've*
> *always pulled together*
> *Who to each other have been true in fair and*
> *stormy weather*
> *Their duty they will not shrink from, now that*
> *they know the worst*
> *So lower the boats – but remember lads, the*
> *women and children first.'*

If one reads almost any of the accounts of the wreck of the *Titanic* in 1912, one is left with a similar impression; the saving of the women and the children was a central part in the drama of the ship sinking. Stories such as these were seized upon as examples of manly courage; of gentlemanly, British and Christian behaviour. Examples of foreigners panicking in similar situations were equally popular. Indeed, one is led to believe from contemporary reports that anyone not British or Christian was not to be trusted in such situations, and that the French were particularly poorly behaved. Lascars, Italians and blacks were also renowned in the British psyche as being less than useless in a shipwreck.

In reality of course the truth was nowhere near so stark. If one digs a little deeper into the stories of the *Birkenhead* and the *Titanic*, for example, it is clear that the men of the *Birkenhead* were coerced into remaining in line; their immobility was not a spontaneous act by any means. At least one account of the sinking of the *Titanic* acknowledged that in isolated cases '*a man here or there momentarily lost his self-command*'. The 'rule' of women and children first was also open to a fair degree of interpretation and in most instances it actually meant the captain's family, followed by ladies, followed by white Christian women and children, and then the rest. In the instances where such a division between men and women seems to have been attempted, it is equally clear that the women being forced to evacuate were not all willing to leave their partners so meekly in the face of death. In reality the victims were all exposed to extreme emotional stress and distress, often in a maelstrom of natural forces in which rational thought made way for instinctive and desperate response.

Nevertheless, these issues dominated the re-telling of the *Droits de L'Homme*, *Birkenhead* and *Titanic* stories, and they are fine examples of how issues concerning politics, race, gender, nationality and class dominated the years in which they sank. But such interpretation is not an isolated phenomenon restricted to those years. Shipwrecks have long been adopted as an important symbol of political, social or religious insecurity and upheaval, and in the same way the very opposite of a wreck – a fully functioning ship – is a long-established symbol of faith and security. Noah's Ark is perhaps the most famous, and also the oldest, example. Consider also the story of Simonides, by the ancient Greek poet Phaedrus. Caught in a raging storm, Simonides' fellow passengers sought to rescue their worldly goods, but Simonides abandoned ship, and left everything behind. Once on shore, his greedy shipmates were robbed by pirates of all they had laboriously saved, while Simonides was loaded with clothes and money by someone who loved his poetry. The Romans were particularly obsessed with shipwreck tales, since death by drowning was considered especially disagreeable, because it robbed a man of the due rites of burial and the pious tendance of his grave by surviving relatives. It also augmented his sufferings in the afterlife – the drowned were forced to wait 100 years on the riverbank before Charon would row them across the River Styx.

Acts of God and Human Error

So often with shipwrecks, however, it is the story of the wrecking itself that is as alluring as the human response to it. How did the *Mary Rose*, the pride of Henry VIII's fleet, the pride of the navy, the pride of Britain, one of the most powerful weapons and most prestigious symbols in the realm, with an excellent sailing record in the 35 years since she had been launched, manage to sink within sight of land in the calm waters of the Solent? How did the *Vasa*, the flagship of King Gustavus II Adolphus of Sweden (r. 1617–32) meet the same fate, this time on her maiden voyage in 1628? What was the sequence of events, apparently so random but in fact intricately connected, that led the *Titanic*, a ship which the press claimed was 'unsinkable', to sink with appalling loss of life? Why was the *Lusitania* full of civilians in 1915 when clear warnings had been given by both the British and American governments that the Germans were preparing to launch a campaign of unrestricted submarine warfare against all shipping in the war zone?

There are of course many reasons – far more than one might assume – for a ship to be wrecked. Wrecks were often, apparently, caused by simple things. A ship, for example, strikes an uncharted rock, or in the case of the *Batavia*, a Dutch ship wrecked off the coast of Western Australia in 1629, the crew mistakes surf on a reef for the shine of the moon. But in reality very few can be explained simply. One can say, for example, that the *Titanic* struck an iceberg, but what was it doing in an area populated by icebergs? Why was the iceberg not avoided? Why did just one collision with an iceberg sink such an enormous ship and why did so few survive? A similar example comes from the wreck of the *Wager* in 1740. Her mizzen mast broke and she was driven ashore. But what was she doing so close to land in the first place, and why did she have no carpenter on board to provide some temporary support to the mast and allow the ship to work her way to windward and to safety?

To some ancient mariners, such detailed questions, now the lifeblood of modern research, would seem a complete waste of time. For those early seafarers, it was easy to believe that before their time, man confined himself to his proper element – the Earth. Neptune's realm was sacred, and to trespass on it was to invoke many evils. The course of a ship over the waves was not only an insult to the outraged sea god, but the furrow driven by the keel and the strokes of the oar blades caused him physical discomfort. Moreover, in ancient times the two reasons for going to sea – war and trade – were both manifestations of human greed: it must not be forgotten that for the Ancient Greeks the first of all voyages was made aboard the *Argo*, to seize the Golden Fleece. Divine retribution was only to be expected.

Cynical modern research offers a different view, but one which is equally entertaining. In place of sea monsters and sea gods we have whales, whirlpools, freak waves and tornados. We have drunk captains and crews, incompetent shipwrights and engineers. We have cannon balls, exploding shells, torpedoes, mines and missiles. The infamous wreckers who lured innocent ships onto rocky coasts are perhaps more myth than reality, but in some cases – in Cornwall in particular – there is some scattered evidence for wrecking.

Archaeology and Exploitation

Fortunately, modern science now allows us to investigate the wrecks themselves and ascertain how some of these ships, whose fate is unknown, actually sank. We can also discover far more about these ships than the reason for their sinking. If the underwater conditions are right, ideally with no daylight and in deep water where the temperature is cold and marine animals find it hard to live, wrecks from hundreds if not thousands of years ago can survive in beautiful condition. In shallow and warm water, a wreck buried in mud can also contain human and plant remains; even clothes can survive.

In poorer environmental conditions other items remain almost unaffected. Stone and precious stones are imperceptibly affected by generations on the seabed. Gold, silver, mercury and platinum are also unblemished. It is impossible to consider all of the extraordinary revelations made by scientific analysis of shipwreck artefacts, but one example makes the point very well. The Ulu Burun wreck, which sank off Turkey during the late Bronze Age – that is three and a half thousand years ago – has recently revealed that part of its cargo was made up of pomegranates, whose juice was used as a highly sought-after source of perfume. Terebinth resin and coriander seeds, both aromatics, were also part of the cargo. Such a collection of plant remains from the Bronze Age is unique.

Wrecks, therefore, can be time capsules, and are rare indeed as an archaeological site. The great tombs of the pharaohs buried deep in the Egyptian earth were sealed, and those that were best hidden remained that way until they were discovered by modern archaeologists. But those tombs, fascinating as they are, captured the life and death of the high-born. The ship, however, is full of the day-to-day goods of people from all social strata, and the majority of ships wrecked enjoyed a long life before their final demise. Indeed the only archaeological site comparable with a well-preserved and undisturbed shipwreck is that of Pompeii, where in AD 79 a town full of men, women, children, merchants, slaves, politicians, soldiers and sailors was smothered in an instant by the eruption of Vesuvius.

On land there are few sites to rival Pompeii, but at sea there is no limit. We know of some stunning examples: the *Mary Rose*, *Vasa*, *Titanic* and the *Hunley* are the best known, but these have been produced by a discipline that is only recently 50 years old: the aqualung was only invented in 1943 and the first proper underwater excavation undertaken in 1960. There are thousands, if not hundreds of thousands more wrecks to find. There is indirect evidence that humans have been travelling at sea for at least 60,000 years and a vast number of the millions of craft that have been built were wrecked. In the Mediterranean in 1992, 1189 wrecks pre-dating 1500 were known. Since then hundreds more have been found. It is no fantastical claim that the seafloor of the Mediterranean is the last great repository of works of art from classical antiquity. In other vast areas of great maritime activity like the Indian Ocean, maritime archaeology is still in its infancy and only a tiny fraction of the real amount of wrecks is known. We have hardly begun looking in places like the Black Sea where environmental conditions for survival are perfect, but whose waters are deep.

All is not rosy, however. The wrecks in deep water are largely free from interference, but those in shallow water have little protection from sport divers, dredging, the laying of oil pipes and breakwaters, and the cables of fishing nets that drag along the seafloor, cutting everything in their path. This is particularly problematic as fish congregate over wrecks; indeed hulks are now sunk deliberately to attract fish and thus provide a focus for sport divers. The fact that marble, bronze, precious stones, silver and gold all survive underwater also lures treasure hunters to wreck sites, some of whom have little or no archaeological training, and material precious to the historian and archaeologist is destroyed or removed from its original context in the search for booty. Such treasure hunting is not a fantastic dream, but a stark reality.

The stakes are astonishingly high and for many the motivation desperate. In most cases in the search for treasure ships, vast sums of money are raised to equip an expedition with the necessary equipment and expertise to locate, excavate and raise a sunken cargo. Every failed dive or blank survey result merely encourages the hunters further, for it is simply proof that they must look elsewhere, and must raise the money for one more expedition. It is in every case much like a gambling addiction: if you lose, then double the stakes, for the next horse is bound to come in. But if indeed there is success, then the rewards are quite astounding. Boxes full of emeralds the size of golf balls; handfuls of pearls; chests of money; bars of gold scattered on the seafloor like so many bricks from a derelict wall. The most famous of all such wrecks, that of the *Atocha*, a Spanish ship laden with treasure from South

America that sank in 1622 off the Florida Keys, made her discoverer, Mel Fisher, a retired chicken farmer from California, an estimated $50 million.

It is more than a little surprising that such ransacking of the world's historical treasure is tolerated, for on land it most certainly is not: the days of grave robbers looting the Valley of the Kings is long gone, and one would be arrested for approaching Stonehenge with anything resembling a spade. But at sea only a tiny percentage of these wrecks are protected and there is no international agreement on how, if at all, such protection should be imposed or policed. Some sunken warships are designated as war graves, but this courtesy is not extended to the thousands of merchantmen sunk in combat. Some high-profile ships such as the *Titanic* and the *Estonia* are subject to multilateral conventions to protect them, but they are not signed by all significant maritime powers. Russia, for example, has not signed the agreement to protect the *Titanic*, and Russian submarines make regular tourist trips to her wreck site, with inevitable consequences for the sustainability of the wreck.

The question of who owns a salvaged wreck is equally complicated. The salvor claims a percentage of the value from the owner, but one must first identify the owner. Moreover, one must distinguish between the ownership of the ship and its cargo. In some cases it is fairly straightforward, particularly so with warships as their cargo and the ship itself is usually owned by the nation in question. Merchant ships are more complex as, in the case of East Indiamen, the ships were usually owned by a syndicate of financiers, and the cargo by the East India Company. Valuable cargos, moreover, were usually insured, and if any insurance claim had been paid out, then the ownership passed on to the insurance company.

The provenance of the cargo itself is also potentially problematic. The wreck of the French warship *L'Orient* is a case in point. In 1798, Napoleon's armies rampaged across Europe, and Napoleon had his sights set on the conquest of Egypt. On the way there, his warships stopped off at Malta, a fortress island run privately and independently by the Knights of St John since 1530. The riches of Malta were renowned, but heavily protected, not least by the Maltese navy, a powerful fleet of galleys that patrolled the straits between Malta and North Africa, keeping the seas clean of Barbary pirates in the name of Christendom. Napoleon arrived in force and begged an audience with the Knights. He was admitted into the hallowed confines of the magnificent harbour of Valletta, when he promptly turned the guns of his fleet onto the fortress. The Maltese had no choice but to surrender, and the island fell. It is rumoured that Napoleon then loaded the riches of Malta onto his flagship *L'Orient* and set off for Egypt, but on 1 August at Aboukir Bay, off the mouth of the Nile, he was caught by Nelson. In one of the most stunning victories in the Age of Sail, the French fleet was annihilated, and the *L'Orient* blew up. Salvors have hunted for her remains ever since, and artefacts began to be recovered in the 1980s. But who owns her cargo? It is a French ship, in Egyptian territorial waters, with a looted Maltese cargo, that is claimed by its modern salvor: it is no surprise that such salvage operations stagnate for years in court and in many cases the salvage is simply abandoned.

Wreck Hotspots

The international nature of shipwrecks and the potential value of their salvaged cargo will always complicate the question of salvage, but it is certain that such cases will only become more frequent in the future, for not only do we now have the ability to see further and with more accuracy underwater than ever before, we can even search remotely underneath the seabed for signs of wrecks. Advances in historical research have also given us a much better idea of where to look, which has in turn been made possible by our increasing knowledge of historic and hazardous shipping routes.

Everyone has heard of the horrors of Cape Horn, where the Atlantic and Pacific Oceans meet in a maelstrom of freezing wind that can blow a ship over, waves that can hide the sun, and currents that can drag a ship with irresistible force to a fate unknown. But less well known is the *mistral* of Provence, a sudden blast of freezing air channelled from the Massif Central through the Rhône Gap and into the Mediterranean. Similar phenomena are the *tramatona* of Liguria, when the wind comes from the Alps, or the *bora* of the Dalmatian coast. Indeed, in some parts of the Mediterranean in winter it is not unheard of for the wind to blow from several directions at once. The Bahamas are also renowned for sudden and violent storms that rise as quickly as they disappear. Fog plagues the coasts of Nova Scotia, San Francisco and Brittany. Low-lying islands in well-travelled trade routes, such as the Isles of Scilly in the mouth of the English Channel, or the Comoros Islands, which lie in the middle of the Mozambique Channel, are both a navigational hazard and a lair for pirates. Narrow straits such as the Bosphorus in Turkey, the Straits of Gibraltar or the Straits of Hormuz that guard the entrance to the Persian Gulf, are renowned for strong currents, and shipping can easily be targeted in these locations from fortified shore positions. Other areas are plagued by ever-shifting sand banks such as the Goodwin Sands off the Kent coast, or the mud banks in the mouth of the Ganges that make access by ship to Calcutta something of a lottery. Ports that are steeped in history inevitably have more than their fair share of shipwrecks. What, we must wonder, would the landscape look like if we could for a brief few hours drain the harbours and outlying waters of ports such as Havana, Vera Cruz, Lisbon or Cadiz?

These 'hotspots' for wrecks might seem quite specific, but in practice they are simply a starting point. The sheer scale of the seafloor is bewildering, and anyone who has dived, or even snorkled over a wreck or a specific area of reef in open sea will know how difficult it can be to find the same spot the next day. To know, therefore, that a ship went down 'off the Florida Keys' is almost meaningless in terms of finding a wreck unless one can narrow it down further. This is where the hours of historical research come in for anyone seeking to find a specific wreck.

We are fortunate in Britain that so much survives in the pages of *Lloyd's List*, a shipping newspaper published regularly since 1734. In Britain the National Archives contain thousands of original ships' logs, and in many cases these are augmented by official enquiries into the loss of a ship or the minutes of courts martial of the captain, officers or members of the crew. In Britain much of this is meticulously catalogued, but archives elsewhere are not so accessible, nor are they so complete. The terrible 1755 earthquake and ensuing fire in Lisbon, for example, destroyed large quantities of records concerning the growth of Iberian seafaring from the 15th century onwards. In addition to these 'official' sources, there are newspaper reports and letters from survivors, some of whom wrote substantial narratives of their experience. There are also ship plans and contemporary charts that provide clues about shipwrecks. No one, however, knows more about the location of wrecks than local fishermen. The sponge and pearl divers who actually spend a great deal of time on the seabed itself know more than anyone. Hundreds of wrecks, particularly in the Mediterranean, have been found by plying these men with local brandy until their tongues loosen.

Given a particular area within which to search, modern equipment can measure changes in magnetic resistivity, which identifies large objects made of iron such as anchors and cannon. Echo sounders and side-scan sonars can measure undulations in the height of the seabed to indicate the presence of a wreck, and sub-bottom profilers can locate buried objects. Those locations can then be fixed by satellite anywhere in the world.

Once the wrecks have been located, the ponderous wheels of academia begin to roll; the remains can then be excavated, analysed and conserved for future generations to enjoy. Hulls and rigging can be rebuilt using ancient techniques, and ultimately the magic of modern technology and research can combine to re-create these vessels as replicas of varying shades of authenticity. If one stops to appreciate what can be done, it is indeed remarkable that we can now sail an almost exact replica of the Kyrenia ship, a ship that was built before Alexander the Great was born. Therein lies the true value of the story of shipwrecks. It is easy to think of that story as one of failure; of bad design and construction, of poor seamanship and incompetence.

But in reality its message is far more hopeful and reassuring: it is a story of enterprise and calculated risk. The shipwrecks are the inevitable wastage of creation; they are the chips of wood and discarded timbers that lie on the shipwright's floor under the hull of a beautifully crafted vessel. To immerse oneself in the story of shipwrecks, then, is to appreciate the continuing story of seafaring, and to bask in the warm glow of human achievement.

Ancient and Dark Age Wrecks

We know surprisingly little about ancient shipwrecks. Because of the rarity of written sources from this period, our knowledge of ancient wrecks relies heavily upon archaeology, and yet we are not fortunate enough to have discovered well-preserved and informative wrecks from every significant age of seafaring, and from every significant seafaring nation.

It may seem surprising to those who have seen images of modern reconstructions of Greek and Roman galleys, but we have not yet discovered the remains of *any* Roman or Greek warships. In fact only one bit of one Greek warship has been found – a cast-bronze ram, known as the Athlit ram, which probably came from a four-banked Cypriot warship around 204–164 BC. From the shape, dimensions and fixing points of this ram we have been able to recreate an approximation of how the bow section of a Greek warship may have appeared, and we can deduce some things about the way that the ram was attached to the hull, but beyond that we know very little indeed about Roman and Greek warships from their physical remains.

Similarly, it is not widely known that modern archaeologists have only ever identified with certainty the wrecked remains of one large 'Viking' warship of the type that transported warriors on raiding parties around northern Europe for 300 years from the ninth century, and that ship, known as Skuldelev 2, actually appears to have been built in the British Isles, probably in Dublin. Those 'Viking' ships that are best known come from rich boat-burials such as Gokstad and Oseberg in Norway, and they are more akin to ornate funerary barges or royal yachts loaded with archaic features, than actual warships with the working fixtures and fittings of a mobile invasion force. The Vikings were also frequently opposed on their raiding parties, particularly by the navy of the early Anglo-Saxon kings of England, and yet we have no physical evidence at all of other European warship types of the same period.

Almost everything we do know about vessels whose physical remains have not been discovered comes from carvings in stone or wood, depictions on pottery, or descriptions in contemporary texts. Those illustrations, however, are often vague, contradictory and could easily have been made by artists with little or no knowledge of ships or shipping. Even if one can determine the age of the vessel depicted it is then difficult to determine if the ship depicted was contemporary with the artist, or, indeed, if he was depicting an old type of ship, or even an imaginary one. For some ancient seafaring cultures, we do not even enjoy the luxury of such difficulties.

There are, therefore, significant gaps within the framework of our archaeological knowledge of shipwrecks, but that framework itself provides a purely arbitrary starting-point. The earliest known shipwreck is the Ulu Burun wreck, from the south coast of Turkey, which dates from 1400 BC, but we know that the first migrations were made between Southeast Asia and Australia in 50,000 BC. Our patchy knowledge of shipwrecks is limited to only the most recent 3408 years of humankind's 50,000-year heritage of seafaring – that is a fraction over 6.8 percent.

Nevertheless, a number of threads unite seafarers even now with the most ancient in man's history; possibly even, with *Homo erectus*, the ancestor of modern man, who modern scholars now believe may even have ventured to sea as early as 200,000 years ago. Seagoing craft sink because they are poorly designed or because they become damaged through storm, warfare, collision or other accident. We also know with some confidence that distinctive stretches of coastline would have formed the basis of navigation. One needs only to view the land from the sea a handful of times to realize how distinctive the coastline can be, and where no distinctive landmarks existed, they were constructed from the earliest of times, and entered the written record around 300 BC. Not long after, sometime between 283 and 277 BC, the most famous of all ancient landmarks, the Pharos lighthouse, was built on the island that protects Alexandria harbour. From its base to the great statue of Poseidon that some believed adorned its peak, the lighthouse rose over 150 metres (500 ft), one of the largest constructions in the ancient world, only surpassed by the Great Pyramid of Khufu at Giza. Another wonder of the ancient world, the Colossus of Rhodes – a giant statue of the sun god Helios completed in 282 BC – performed a similar function for mariners entering Rhodes' Mandraki harbour. Lighthouses spread with the geographical reach and engineering genius of the Roman empire and the remains of many litter the most significant headlands of Europe. A partially intact and fine example of a Roman lighthouse can still be seen today at Dover on the south coast of England. Norway's southern coast boasted a fine collection of Viking seamarks, some of which still survive.

Ancient Methods of Navigation

When in open sea, the sky is full of information useful to mariners. Navigation would have been based on the predictable daily movement of the sun from east to west, and the rising and setting of the stars: once one is oriented and possessed with a desired direction, if not a location, then a voyage becomes possible. Land itself is often signalled by a stationary cloud in an otherwise clear sky, or in a sky where other clouds are moving. In low latitudes the underside of a cloud near land might have a greenish-blue tint, and in high latitudes a light tint. Bird flight at dawn suggests seabirds are flying from their nests to a feeding ground at sea, where they return at dusk. The presence of flies, butterflies, bees and other insects suggests the presence of land, from where they have been blown by the wind. We also know that land-sighting birds were also used for navigation. The earliest description comes from the Bible, in which Noah releases doves to find land. Pigeons were also used, and the Vikings favoured ravens.

The wind itself can suggest land, as in certain latitudes the wind blows from the land at midnight until midday, and then reverses its direction and becomes an offshore breeze. This is caused by changes to the temperature of the sea and land as they are heated up by the sun, and so therefore is normally associated with hot environments such as the Mediterranean summer. In some latitudes at certain times of year, winds are so regular that they even became named after the country from which, or the country to which, they blew. In ancient Greek the word 'wind' meant very much the same as 'direction'. Specific routes and specific winds were therefore well known. Thus

Pliny the Elder (23–79), a Roman philosopher, historian and naval commander, defined the French wind called Circius as the one that would blow you from Narbonne (on the south coast of France) across the Ligurian Sea to Ostia (the port at the mouth of the Tiber). Navigation in the lands of these predictable winds was far easier than elsewhere.

Even smell was occasionally useful for navigation. Certain continents carry very noticeable odours which carry for miles at sea; the distinctive smells of peat from Ireland or eucalyptus from Australia are perhaps the best examples of this. The surf pounding on a beach or an offshore reef can be heard for miles at sea. The colour of the sea can also change when land is approached, noticeably lightening as it becomes shallower. A sweet taste or muddy hue of the water suggests a river discharging into the ocean nearby. In northern latitudes seal and sea-lions are only seen close to shore. With the exception of the Sargasso Sea, over 5 million square kilometres (2 million sq mi) of the mid-Atlantic that is covered in huge mats of Sargasso seaweed, driftwood, vegetation, kelp and other flotsam were often taken as indicators of land.

Undoubtedly, however, it was the sea floor that was as useful as the sky in ancient navigation. Sticks or weights attached to twine could be used to measure depth, and also to navigate. To be in soundings – that is in depths of less than 100 fathoms (183 metres/600 ft) – is the first sign that land is close, and that the vessel has crossed the continental shelf. A rapid decrease in depth then is proof that land is approaching fast. The make-up of the sea floor itself was also crucial as it is distinctive in different locations. It may be thick with weed, or covered with fine or coarse, yellow or red, black or white sand; it may be black mud or rock, pebbles or shells, even boulders. To ancient navigators these were as well known and as certain clues to navigation as promontories and inlets, capes and bays, hills, mountains and volcanoes. Indeed, the sounding pole is the earliest known navigational instrument, and is clearly depicted in numerous Egyptian paintings from 3000 BC.

To secure themselves from dangerous winds and currents that might blow them on shore, ancient seafarers used anchors in exactly the same way we do now. We know from some wreck sites that an ocean-going ship would carry several anchors, usually made out of stone. These were particularly good for anchoring on a stone seabed. Stone anchors with holes in, in which wooden stakes were driven, have also been discovered. These were probably used to secure a ship in a sandy or muddy bottom, the stakes digging into the mud and holding the ship's weight. The earliest anchor cables were made of hide and animal membranes, and later of flax, jute, hemp and cotton. The advantage in holding power of a chain cable over a buoyant cable was recognized quite early on, and we know that the Greeks used a few fathoms of chain on their anchor cables from 500 BC, and the Chinese possibly earlier. There is also detailed evidence in Caesar's journals of a French tribe from southern Brittany using anchor chain in the sixth century BC, and we also know that the Vikings used anchor chain from the ninth century. The use of anchor chain then disappears from history for over 1000 years.

By a combination of these means the Egyptians successfully navigated from 3500 BC onwards; the Minoans traded in the Mediterranean from 2700 BC; the Polynesians migrated deep into the Pacific around 1000 BC, and at roughly the same time a Phoenician seaborne trading empire began to dominate the Mediterranean. Some believe they ventured well beyond the Pillars of Hercules and that they even visited Britain.

By the fourth century BC early sailing directions in a form of pilot-book begin to appear and it is in these that local navigational knowledge was written down and passed on. Their use, however, depended entirely on a level of literacy in the user that was not widespread among the seafaring community. For example, educated Roman naval officers may have understood and used them, but the great majority of local and isolated fishermen would not. With the occasional augmentation by detailed pilot-books, therefore, the art of navigation remained largely unchanged until the gradual adoption by mariners of the magnetic compass, astrolabe and sea chart after 1200. But by then Arab merchants had been trading with China for 450 years and it was two centuries since the Vikings had reached North America.

Navigation, therefore, was very much an art form in these years, and the ease with which it could be done varied greatly throughout the world. Although the summit of a mountain some 2500 metres (8000 ft) high can be seen at sea up to 100 miles (160 km) away on a clear day, for example, there are no such heights in Egypt or Libya. Nor do the regular winds of the Coromandel Coast of India, or of the eastern Mediterranean, blow further north. Nevertheless, the Vikings took to the great oceans, blustery winds and dark skies of the north with arguably more success than the Greeks, Phoenicians and Romans enjoyed in the Mediterranean, though even they had the special term *hafvilla,* meaning 'lost at sea'.

It is from this period, before the art of navigation became scientific, but in which long ocean voyages aboard large ships were possible; merchant ships with vast cargoes voyaged hundreds of miles on a regular basis; and great battles were fought at sea, that the following four wrecks have been taken. They have not been chosen because they represent an even coverage of time or location: there are insufficient wrecks to choose from to enjoy that luxury. They have, rather, been chosen because they are particularly instructive about the past. Only a handful of known wrecks survive from those years in good condition, and these are the best of them.

The Kyrenia Ship

c.300 BC

The excavation of the wreck of the 'Kyrenia ship' off northern Cyprus in 1967 was a key moment in the science of maritime archaeology, then in its infancy. This vessel – a fourth-century BC Greek merchantman – provided new insights into ancient shipbuilding techniques and shipping practices. The fine state of preservation of the ship and her cargo initially suggested that she had not met a violent end, until further investigations began to hint at a more sinister turn of events.

Kyrenia is a small port on the north coast of Cyprus, the most easterly of the Mediterranean islands. A stepping-stone between three continents: Europe, Asia and Africa, Cyprus has always enjoyed strategic significance for warriors and merchants alike, and has always been a melting-pot of cultures, fashions, ideas and art. Indeed, its artwork has been highly distinctive since prehistory. In the fifth century BC, Cyprus became the buffer between the two warring kingdoms of Greece and Persia, and until it was taken over by Alexander the Great (r. 336–323 BC) in 333 BC, the island remained under Persian control. In these years Cyprus was renowned as a source of copper, corn and timber; even today, almost a fifth of the entire island is forested, with a magnificent variety of coniferous and broadleaved trees, all ideal in their own way for different types of shipbuilding. The third-century BC geographer Eratosthenes claimed that the island was then almost entirely forested. Unsurprisingly with such a rich stock of timber, Cyprus enjoyed a reputation as a shipbuilding centre, particularly around the fine natural harbour of Nea Paphos that was surrounded by rich forests and located on the southwestern corner of the island, where it faced the alluring coast of Egypt. It was in this world that the Kyrenia ship sailed, at an important period in world history when the dominant Persians were being challenged and defeated by Alexander the Great. She was just one of those merchant ships that were the life-blood of the region, linking people and empires across the Aegean in a web of trade, and generating the wealth that made the Mediterranean one of the cradles of civilization.

Until the late 1960s, however, we knew almost nothing about ships and shipping of that era. The scant evidence that was available came from illustrations on vases, eroded carvings on tombstones and ancient writings. But then, in the summer of 1967, the Cypriot government invited an archaeological team to investigate ancient shipwrecks off the coasts of the island and they were first taken to a small pile of amphorae by a sponge diver who had found the wreck some years before, hidden in the dancing Poseidon grass that covered the site.

A Mixed Manifest

It did not at first appear to be a substantial find. Amphorae were the common cargo vessel for liquid, and thousands have survived scattered across the Mediterranean Sea bed. Nevertheless, the archaeologists recorded the pile of amphorae and noted that they were remarkably well stacked, as if the ship had carefully sunk to the bottom, cradling her cargo. There were no reefs or isolated jagged rocks nearby, further suggesting that the last moments of this ship were not violent. Moreover, it only took a brief investigation to realize that the visible pile of amphorae was only part of the wreck. Rather like an iceberg, it appeared that there was far more buried in those Cypriot sands. The archaeologists determined to excavate. It must be remembered, however, that in the mid-1960s the science of working on the sea floor was still relatively young, particularly so at depths of 30 metres (100 ft) where the Kyrenia ship lay. The time that divers spent submerged was limited to prevent them from getting the 'bends' (decompression sickness, a potentially fatal condition caused by nitrogen bubbles in the blood), and a recompression chamber was fitted on the barge from which they dived, so that anyone badly affected could be treated immediately. All of this seems expected and commonplace to us now, but this was one of the very first full-scale archaeological investigations at such a depth, and it became a benchmark. To aid communication, a large container the size and shape of a telephone booth was sunk and filled with air from the waist up. This allowed divers to enter the booth and hold a telephone conversation with the support barge above while also serving as an emergency shelter if a diver's breathing equipment failed. The systematic approach was rewarding, and what they found astonishing. A few feet under the sand and mud lay the preserved hull of a Greek merchant ship dating from the fourth century BC. Sixty percent of her hull still survives, more than any other ship from the ancient world.

Later analysis showed that her cargo contained 11 different types of amphorae, 400 in all, which seemed to have been neatly divided into two separate cargoes. The cargo amidships consisted of amphorae all of similar size and shape, coated on the inside with a thick layer of bitumen or resin. This kept them watertight for the transport of liquids such as wine, water or oil. At either end of the ship, amphorae of a noticeably different shape, indicating a separate consignment. The main cargo, 343 of the 400 amphorae, was identified as coming from Rhodes, an island famed throughout the classical world for the quality of its wine.

Below the amphorae lay huge volcanic grinding stones, possibly used for grinding grain into flour, but experiments using the blocks produced very poor quality flour and it is likely they were used for something else, perhaps grinding ore. The clear lack of uniformity in the 29 stones discovered also raises the possibility that they had been left behind from a previous consignment and were being used to ballast the ship. It was also clear from the distribution of the surviving cargo that a ship stowed in that way would have been very unbalanced, and it is thought likely that she was also carrying a large cargo of perishable items, possibly bolts of cloth, that have long since disintegrated. Perhaps most surprising, though, was the immense cargo of almonds that the ship carried, nearly 10,000 in total. The nuts themselves had all gone, but their shells remained heaped in the bow. The clear lack of any associated amphorae suggested that these almonds had been transported in sacks, now rotted away. It is most likely that these had been recently loaded in Cyprus, which was renowned for its almond orchards in antiquity, and still is.

Away from the main cargo the excavators came across much finer pottery, used for cooking and eating. There were drinking cups, ladles, sieves, a fragment of a lamp, five bronze coins, even an ink well. The dishes, cups, oil jars and spoons discovered were all identical, and in each case there were four of them, which clearly suggested the number of the crew. Nearby, they also found the remains of food: olives, pistachios, hazelnuts, lentils, garlic, sprigs of dried herb, grapes and figs. The cooking utensils and food remains were found in the stern, but all of the cups were found together in the bow, which is where, in the traditional Mediterranean caiques, the drinking water is still kept. Analysis of this pottery showed that it was all made in Rhodes, like the main cargo of amphorae. It is considered likely, therefore, that Rhodes was her home port.

New Light on Ancient Shipbuilding

The hull itself had clearly not been sound at the time of the wreck, and there were visible signs of repair and further evidence that the ship had enjoyed an extremely long life. Indeed, radiocarbon dating of the ship's hull showed that she was at least 80 years old when she sank, and it is quite possible that she could have been significantly older than that. It was clear that one of the ship's frames had at some point been replaced, possibly not long before the ship sank, for it was made and fastened in a different way from the others and had no signs of the wear and tear that years of shifting cargo had inflicted on the ship's other internal timbers. The bows had also been sheathed in pine before the entire hull had been sheathed in lead to prevent the *Teredo navalis* (the shipworm) from further weakening an already ageing hull. The lead had been tacked onto the hull with copper nails on top of a layer of agave leaves, woven and saturated in a resinous pitch. A spare roll of lead and a small wooden mallet were found in the bows, which presumably acted as a sort of puncture-repair kit.

In the final analysis, the Kyrenia ship was evidently Greek, and dated from the fourth century BC. The oldest of the five coins was from the reign of Antigonus I Monophthalmus ('the one-eyed'; r. 316–301 BC), a successor to the empire of Alexander, but the timber for the hull was felled 44 years either side of 389 BC. It is impossible to be more precise, but we can say with confidence that she was built before Alexander the Great was born, and that she sank around 20 years after his death.

A Victim of Piracy?

But what of the sinking itself? This has always been a bit of a mystery. The Cypriot climate is steady in the summer, and it is unlikely that the Kyrenia ship would have braved winter storms. Indeed, we know that seafaring in the ancient Mediterranean was carried out almost exclusively between April and September, when the breezes are light and predictable, and the sea calm. The Mediterranean adopts an altogether more unpredictable and violent nature between late autumn and early spring. Nevertheless, on very rare occasions, Cyprus can be vulnerable to winds that scream down from the Taurus Mountains, and the complete absence of any mast at the wreck site suggests that it had been carefully taken down, perhaps to face an oncoming storm. Lead rings found in the stern suggest that the sails might have been stowed there temporarily, as the lead rings would have been used as fairleads and cringles to secure the running rigging to the sails. The lack of a hearth on board and only one tiny fragment of an oil lamp suggest that this timid trader rarely, if ever, sailed at night. At first, therefore, it was considered likely that she was lost in the daytime, at some point in the summer months, perhaps overcome by a sudden squall. But gradually the mystery began to deepen.

The excavators were struck by the small number of coins found on the wreck. Only five survived, and yet one would expect a merchantman such as this to carry a greater purse with which to conduct trade. With the exception of the crew's cooking utensils, there was also a marked absence of personal belongings. One explanation was that the crew had enough time to abandon ship with their most treasured belongings, and perhaps they made it ashore at Kyrenia, only half a mile (0.9 km) distant. When the hull was finally raised, however, new evidence cast doubt on this hypothesis, for underneath the hull lay several heavily corroded iron fragments. Their location suggested that they had been embedded in the hull when the ship sank. When the excavators broke them open, to everyone's surprise they were all iron spearheads. Had this sluggish merchantman been chased down by a swift-oared pirate galley, her crew seized and sold into slavery, their money and personal possessions stolen, and the ship scuttled to destroy evidence of the crime?

It is more than possible. Numerous depictions of Greek ships survive and although the artists tended to favour the slim lines and latent aggression of warships, one fine example survives on a Greek cup in the British Museum, showing a merchantman being pursued and then attacked by pirates. The pirate ship is long and low, powered by both oar and sail with a large crew. She has a formidable-looking ram bow with which she threatens the merchantman. The merchantman, on the other hand, is a much larger, beamier ship, powered by sail alone. The only crew member visible is sitting at the stern, manning a side-mounted steering oar, the precursor of the stern-hung rudder. In the first image the merchantman is proceeding cautiously with a clearly reefed sail; perhaps the winds are strong or the weather threatening. The pirate ship, by contrast, is under full sail and oar. Then, in the second image the situation changes. The merchantman has recognized the threat posed by the galley and has unbrailed her sail in a bid to escape. Meanwhile some of the pirates have stood up from their oars and are taking in their own sail, perhaps in preparation to board the unlucky merchantman.

We do not know how the scene depicted on the vase ended, nor do we know for sure how the Kyrenia ship came to rest on the seabed, but there is no doubt that events such as those shown did take place. Piracy in the ancient Mediterranean was common; it was one further means of livelihood that the sea supported. Wealthy trade plied these waters on predictable routes, most within sight of the barren coastal land which provided poor quality soil for farming. Piracy was an extremely attractive option. There are scattered references to piracy in ancient Greek law, with provisions made for the illegality of their activities in the *Lex Rhodia*, a collection of legal practices relating to commerce and navigation in the ancient Mediterranean. In later years the Romans were quite explicit in how they dealt with pirates: they were beheaded, crucified, or 'exposed to beasts'. (Julius Caesar's great rival in the First Triumvirate, Pompey, made his name campaigning against pirates who threatened the Italian coast in the mid-60s BC.) There is no similar evidence from ancient Greece regarding such punishments apart from a contemporary inscription from the city of Ephesus in Asia Minor, which states that pirates were dealt with 'in a manner that befitted their villainy'. The implicit evidence from the Kyrenia ship is that their activities were villainous indeed, but perhaps they were returned in kind.

The Intan Ship

*c.*920

From as early as the seventh century AD *onwards, there was a thriving seaborne trade in luxury goods both within Southeast Asia and between there and the Chinese mainland. Later, Arab merchants ventured across the Indian Ocean, bringing commodities from the Middle East to this lucrative market. Indeed, this was how Islam spread to China from the ninth century and later to Sumatra and Java. In an area where shipwrecks are often plundered before archaeology can take place, an extraordinary wreck found off Sumatra in 1997 has helped throw important new light on the early maritime civilizations of Southeast Asia.*

We know from ancient Chinese texts that Emperor Taizong (r. 627–49) of the Tang Dynasty frequently suffered from diarrhoea. We also know that once, when he failed to recover from a severe bout using traditional remedies, his doctors mixed milk with 'long pepper' in a final attempt to slow the royal gut. This remedy worked, but what is particularly interesting about this cure is that the pepper it contained came from Indonesia. Moreover, pepper was not the only Indonesian product in demand in China during this period. Pine resin was a miraculous cure for boils and suppurating ulcers. Cloves, costus, the rattan palm, and camphor were all highly prized by Chinese doctors. Indonesian jungles were full of aromatic trees such as sandalwood, and other natural items sought after by a world increasingly caught in the grip of luxury. One of the most desired was *guggulu*, an Indonesian equivalent of myrrh that burned with a powerful aroma, and many believed with a superior perfume.

Early Demand for Luxury Goods

Written records from the court of the succeeding Song Dynasty, around the tenth century, show that 'dragon brain' camphor from Sumatra and the Malay Peninsula was in great demand, along with silk, tortoiseshell, white cardamom, cloves, pepper, nutmeg, sandalwood and kingfisher feathers. From further west, along the coast of what is now southern India, many other items had reached China and were presented by traders as 'tribute'. We know of rhinoceros horns, 'spiral' and other 'medicinal' horns, elephant tusks, pea-fowl, rattan mats, embroidered woven cloth and coconuts. Even peacock-feather umbrellas were presented to the Song court.

We cannot be certain when this trade began, but we do know that by 1000 it was well established. By then Arab traders also carried luxury goods of the Middle East from the port of Basra on the Tigris to China. Their names survive in the court registers of the Song Dynasty. At first they appeared to be curious Chinese names until modern scholars realized that they were Chinese renditions of Arabic names, and the repeated surname 'Li' actually translated as the Muslim 'Ali'. All of those names have since been recreated, and thus Li He-mo becomes Ali Mohammed, and Li Mu-zha-duo becomes Ali Muzaffar.

These were the treasures of the East, the luxuries that would enrapture Western populations as Portuguese, Dutch and English merchantmen started to become involved in the trade from the 15th century onwards. But this was at least 500 years before that Western intrusion – the period when the Vikings dominated northern Europe and the luxuries of the East only found their way to Western societies in tiny quantities. On the other side of the world, however, this was the first great age of Southeast Asian commerce, when trade flowed from the east coast of Africa, the Persian Gulf and India through the archipelagos of Indonesia and on to China. Ports involved in that trade became colourful melting-pots, cosmopolitan urban centres of a size and sophistication such as the world had never seen. We know from written sources that the southern Chinese port of Guangzhou (Canton) had a Muslim community from as early as 851. By then, Hindu communities from India had thrived in both China and Indonesia for almost a century.

Dominating the Seaways

The great economic success that followed the upsurge in productivity in China led to population growth, which in turn generated more demand. The vast increase in the opportunity to generate wealth that ensued led to competition for control of the seas in which the merchantmen sailed. The route from the Indian Ocean to China is blocked by Indonesia, and there are only two easily navigable straits through the islands. The first and most northerly is the Strait of Malacca, which lies between the southwestern shore of the Malay Peninsula and Sumatra. The other, further to the south and therefore representing a longer journey, is the Strait of Sunda, which lies between the islands of Sumatra and Java. Both are very narrow; the Sunda Strait is only 15 miles (24 km) at its narrowest, while the Malacca Strait is a mere 1.5 miles (2.4 km) wide, near modern Singapore.

These straits quickly became the focus of a power struggle to control this lucrative trade. Out of that competition arose the great kingdom of Śrivijaya, which controlled the

Malay Peninsula, Sumatra and Java – and hence both of the narrow straits. Śrivijaya was, more than anything, a convenient place to revictual on the way to or from China, and to wait for the appropriate monsoon winds to blow. As the mid-point between China and India, it was also the most convenient place to exchange goods.

The origins of Śrivijaya are obscure. Only a handful of engravings in old Malay and the writings of a seventh-century Chinese pilgrim, I Tsing, who visited Śrivijaya in 671, make any reference to its origins at all. All we know is that its capital was Palembang and that it had established itself as the foremost commercial power in western Indonesia by the end of the seventh century. There should, it follows, be a great deal of physical evidence directly relating to the ships and shipping of this period in and around the coasts of modern Sumatra and Java. No doubt there is, but maritime archaeology itself is a very young discipline and its roots lie firmly in the Mediterranean. Only in the last decade or so have professional archaeologists begun to explore the treasures of Southeast Asia, while government support for archaeological excavation there is minimal.

Valuable Finds

The clear abundance of shipwrecks in those crystal-clear waters, moreover, has made them attractive to treasure hunters keen to exploit the remarkable state of preservation of Chinese porcelain, which in many instances survives intact, in vast quantities, and fetches astronomical sums at auction. Chinese blue and white porcelain from the *Geldermalsen* wreck recently sold for $15 million at auction. Those wrecks that are discovered and looted in this way are not recorded to acceptable professional standards, and material of historical value is destroyed. The Vung Tau wreck, for example, is the only example ever discovered of a *lorcha* – a ship containing elements of both Western and Eastern shipbuilding practices together. And yet her hull was torn apart by salvage divers to get at the valuable ceramics inside. It is a terrible irony, therefore, that this part of the Earth's seabed, which contains as many wrecks of historical importance in as great a density as the Mediterranean, has yielded so few that advance our knowledge of history. Those that have been professionally excavated and recorded shine like gemstones through the murky past. One of those is known as the Intan wreck.

The Intan wreck is one of the most significant wrecks ever found in Southeast Asia. It is the oldest Southeast Asian wreck found with both the cargo and a large portion of the hull intact, and is the first archaeological evidence ever uncovered of shipping in Southeast Asia during the first millennium AD. Since it was discovered in 1997 only one other wreck has been discovered to pre-date it, the

Belitung wreck, an Arab or possibly Indian vessel, which has been dated to *c*.838.

The wreck, which was found by fishermen and named after the nearby Intan oilfield, lies in the open waters of the Java Sea some 45 miles (72 km) off the coast of Sumatra. It was quickly established that it was a site of enormous historical significance and a German company, directed by a professional maritime archaeologist, was contracted by the Indonesian government to excavate it. The most notable feature of the Intan cargo is its diversity. There is physical evidence of beautiful craftsmanship, advanced technology and religion, as well as international trade on a scale corroborating the written sources. Almost no other evidence of artefacts from this era comes from shipwrecks, but a substantial amount of data survives from excavations on land throughout China, Asia and the Middle East that allows us to understand the origins of the ship's cargo. Each item is revealing in its own way and the wealth of data that has been teased from these artefacts by modern research is deeply impressive.

Among the Intan ship's diverse cargo of ceramics, numerous utilitarian brown-ware pots have been identified as originating from Guangdong and Fujian, south-eastern coastal provinces of China. Green-ware bowls, finely decorated boxes and jars probably came from the more northerly Zhejiang province. A number of exceptionally high quality items of white-ware, possibly originating from the famous Xing kilns, were also found. Ceramics of this quality were rare indeed and it has been speculated that they were a gift to royalty. Other ceramic items came from southern Thailand and the Middle East.

A Local Vessel Trading in Scrap and Silver

The ship also carried a large cargo of metal. Thousands of tin ingots probably came from the rich tin mines near Kedah on the Malay Peninsula, and intermingled with the tin was a cargo of lead and also of bronze in bars or distinctive dome shapes. Scattered among the metal cargo were broken fragments of bronze statues. These had not been damaged in the wreck or even in the intervening 1000 years on the seafloor, but had been loaded in this condition – this was a cargo of scrap metal, probably on its way to be melted down and reworked. The sculptures were definitely Indonesian in origin, probably cast in Sumatra, but they showed strong Buddhist and Hindu influences from India. They included religious items, such as a sceptre and bells used by priests as signs of office and to help run their ceremonies, and also temple door-knockers beautifully carved with hideous beasts. A number of bronze mirrors were also discovered, which came from both China and India.

In addition, the ship was laden with over 100 bars of silver, which from their inscriptions were originally payment for a salt tax, but perhaps had now become an illicit cargo, to be melted down and reworked along with the tin and scrap bronze. The gold coins on the wreck were larger than any others ever discovered in Indonesia. They were minted in Sumatra. They were also discovered near a hoard of rings, set with amethysts, rubies, emeralds, sapphires and rock crystals. Nearby lay a cargo of Middle Eastern glass in an array of beautiful colours: clear, pale and dark greens and blues, even purple. There were hundreds of glass beads, each decorated with an eye for protection against evil. Elephant tusks and teeth, carved ivory, deer antlers, earrings, cattle bells, aromatic resin, thousands of candlenuts – still fragrant after over 1000 years on the seabed – tiger bones and teeth, and a few human bones including a highly polished human molar, made up the rest of the artefacts recovered.

To where was this extraordinary cargo bound and why? She was not a Chinese vessel, nor is it likely that she collected her cargo of Chinese porcelain in China, her Indian goods from India, or her Middle Eastern goods from the Persian Gulf. We can deduce this from her construction; built using a local 'lashed-lug' method, she was evidently an Indonesian ship sailing her home waters, and probably picked up her diverse cargo at a nearby 'entrepôt' port. The wreck lies between Sumatra and Java, and since none of the bulk items originate from Java (which is almost devoid of commercially viable metal deposits), and some are certain to have come from Sumatra, it is most likely that the ship was sailing from Sumatra, possibly from Palembang, to Java. It has been suggested that the luxury goods were destined for Chinese envoys, or possibly wealthy Chinese merchants living on Java at that time. It was certainly a troubled time for Java, as the Javanese court was forced to move after a devastating earthquake around 930. It may be that this shipment was en route to supply the newly founded eastern capital with the utilitarian items and raw materials needed to rebuild its shattered community.

A Shift of Focus

The Intan wreck helps to fill an important gap in our knowledge of this era, but so many gaps still remain for this geographical area, and also for this period. The Intan wreck was one of those Indonesian and Arabic ships that we know traded in Chinese goods, but we also know that between the eighth and 12th centuries the Chinese developed the large ocean-going junk that was to protect trade and send powerful armies on both invasions and lengthy voyages in the 13th, 14th and 15th centuries. As yet, however, we have not discovered a single Chinese wreck that pre-dates 1271, the date of the Quanzhou wreck, a Chinese ship laden with trade goods from South-east Asia. Our knowledge of wrecks in this area is also heavily biased towards those with ceramics as cargos, and there is a rather depressing reason for this. Shipwrecks here are discovered by fishermen who bring up artefacts in their nets or snag those nets on wreckage. If they recover wood or perhaps resin, the artefact is usually thrown back, but every Indonesian fisherman is aware that money can be made from ancient ceramics. There are few other ways to discover wrecks in Southeast Asia as there are no detailed archives of shipping to consult. Even so, the focus of maritime archaeology is slowly shifting to this region, and in the years to come archaeologists will recover from these waters artefacts that will change the way we understand our world. It is one of the most tantalizing promises of the 21st century.

The Skuldelev Ships

*c.*1000

The Vikings, whose voyages took them as far afield as Newfoundland and Constantinople, were undisputed masters of the northern seas for almost three centuries from the late 700s onwards. Viking vessels are commonly thought of simply as 'longships', but in 1957, a discovery in Roskildefjord in Denmark revealed five Viking ships of widely divergent types. Built to the highest standards of craftsmanship, the Skuldelev ships bear witness to a highly sophisticated seafaring culture, and stand as the definitive engineering achievements of an era and an empire.

In 1016 Cnut the Great of Denmark (Canute I; r. 1016–35) became the undisputed ruler of England. He was not the first Dane to be king of England; only three years earlier Svein Forkbeard had enjoyed a short reign, curtailed by his death. But by 1016 the Danish influx was irresistible. They had been raiding British shores with great success for more than two and a quarter centuries. In 789 the very first Vikings had arrived at Portsmouth unheralded, and had made their intentions clear by slaughtering the local official who met the landing party on the beach. Four years later the island monastery of Lindisfarne, off the Northumbrian coast, was sacked. Christian Europe trembled under the onslaught of these pagan raiders, who chose to attack on Christian feast days when they knew that there would be little resistance.

Presently, attacks on Britain became attacks on northern Europe, and coastal attacks became attacks deep inland, as the Vikings used their fast and shallow-bottomed warships to penetrate the soft underbelly of their rich neighbours. In 845 they made it to the gates of Paris. By 865 the raids had become invasions: this new generation of Vikings was intent on conquest, not plunder. In the British Isles, Dublin, York and the Isle of Man became Viking kingdoms in their own right. That same quest for land also lured them across the Atlantic, where they had colonized Iceland by 930, Greenland by 936, and had landed in North America (at L'Anse aux Meadows on Newfoundland) by 1000.

Britain's Anglo-Saxon kings resisted the Vikings. The exploits of Alfred the Great, king of Wessex (r. 871–99) are well known, but less so are those of his son, Edward the Elder (r. 899–924), who we know in 910 trapped a Danish force with a fleet of 100 ships, the first English fleet of any size known to have existed. In 936, Edward's son Aethelstan (r. 924–39) then fought a great battle at Brunanburh, where some sources say he faced 615 Viking ships. King Edgar (r. 959–75) did much to strengthen the English navy and his successor Aethelred II (the 'Unready'; r. 978–1016), used it to raid the Vikings on Man. But he was unable to resist growing Viking attacks. After 991 they were repeatedly bought off with Danegeld – vast sums of money paid to ensure the safety of English towns and cities. But, as with most protection rackets, the ability to pay a heavy ransom was a dual acknowledgement of weakness and wealth, the meat and drink of the Viking world, and they soon tired of raiding and sought to conquer. With only a brief intermission after Svein Forkbeard's death, when Aethelred reclaimed the throne in 1014, Danish Vikings ruled England for almost 30 years.

Five Diverse Vessels

It was at the height of Viking power that five ships (the Skuldelev ships, numbered 1–6, with no number 4; when Skuldelev 2 was excavated, she was first thought to be two ships) were beautifully crafted from oak, pine, lime, birch and alder; their planks specially chosen so that the natural curve of the grain mimicked that intended for the ships, thus ensuring that the vessels could be built with fewer joints. This gave them greater strength and resilience.

Every oak plank had been fashioned from a trunk at least 1 metre (3.3 ft) in diameter, the log split into 32 sections of uniform width. Cut in this way, the planks were light but exceptionally strong, and the wood almost shrink-proof. The planks themselves were tenderly cut with an iron axe and a plane; not one had been sawn. It was ships like these that enabled the Vikings to stretch their distinctive culture further than any other European nation until the advent of Columbus and the Iberian explorers in the late 15th century.

Until 1957, however, our knowledge of Viking ships was limited. It derived mainly from the descriptions of ships and shipbuilding in Norse poetry, as well as from their depiction on tombstones and the physical remains of ships in a number of ship burials. But a blockage at Skuldelev in Roskildefjord on the Danish island of Zealand was soon to answer a great many questions. The blockage was a well-known hazard to shipping that lay across the most direct channel from the fjord to the town of Roskilde itself. It had been known for some time that there was a sunken ship there, but it was believed to date from *c.*1400 and was known as 'Queen Margarethe's Ship' after the Danish queen who unified Scandinavia around that time. In 1956 an amateur diver recovered a section of the wreck and took it to the Danish National Museum, where it was shown to be Viking, and was at least 400 years older than first believed. Not only that, but subsequent investigation proved that the blockage was not caused by a single wreck, but by five ships; two warships and three merchant ships. They were all Viking, and most importantly, each one was of a different type, size and design. It was a discovery that electrified the nation. Much as the rediscovery of the Tudor *Mary Rose* became something of a national talisman in England in the 1980s, so were these Viking ships at once cherished and championed by the Danes. A war chest was raised to create a cofferdam around the site, sheltering a full 1579 square metres (17,000 sq ft) of the fjord's floor. Thus exposed, the ships could quickly be recorded and recovered.

It was almost immediately clear that the five ships had all been deliberately sunk. They had been positioned carefully at the mouth of the fjord and laden with stones. Of the five, four had no close parallels in any other known Viking ship. The one that did, a small warship, was almost a perfect copy of a vessel found in a ship burial at Ladby in 1935, but at Ladby no actual timber had survived, only her imprint remaining in an earthen barrow, scattered with rusty nails: until now, archaeologists of this type of ship had only been able to study shadows. Archaeologists, moreover, had been deeply sceptical about the seaworthiness of the Ladby ship and some considered her to have been built purely for burial. The existence of a similar ship, which at some point in its life had clearly been seaworthy, transformed their opinion and rewrote our understanding of Viking amphibious operations. This warship was exceptionally small, narrow and light, a little over 17 metres (56 ft) long and only 2.6 metres (8.5 ft) wide – that is seven times as long as she was wide – and just over 1 metre (3.3 ft) high amidships. She would be lightning fast under sail or oar and could carry 24 men, who could easily haul her onto a beach, a distinctive characteristic of these ships. Indeed the scars and markings on the bottom of her hull showed that this had happened countless times in her evidently long life. Once hauled ashore any horses that had been embarked could easily step off, straight onto dry land, a huge advantage in amphibious operations. Fully manned, she had a draught of less than a metre (3 ft), and when a replica (the *Helge Ask*) was built after the excavation she rode the waves lightly and carried her sail well.

The other warship from Skuldelev was an ocean-going longship. She is a full 12 metres (39 ft) longer than the smaller warship and twice as wide. She probably carried 20 25 pairs of oars, and a total complement of as many as 80 men. Her remains are few and poorly preserved but she is the only undoubted large Viking warship that has ever been excavated. The timber for her construction was felled in 1040; she was built in Dublin in around 1042 and repaired in the 1060s. It is now believed that she spent those early years of her life in and around the Irish Sea, and it is possible that her journey to Roskilde was in response to William's invasion of Britain in 1066. After Harold was defeated at the Battle of Hastings, two of his sons and a daughter fled to Dublin. From there they launched attacks on Norman England in 1069, co-ordinated with attacks led by King Sven of Denmark. The year before those attacks took place, a diplomatic mission was sent from Dublin to Denmark to establish an alliance that would work to restore Anglo-Danish rule in England. This large warship may have formed part of that diplomatic mission. The Viking Ship Museum in Roskilde that houses the Skuldelev ships built a full-scale replica of this ship between 2000 and 2004. Christened *Sea Stallion from Glendalough*, she took around 40,000 man-hours to build and cost over 10 million Danish crowns. Some 150 cubic metres (5296 cu ft) of wood were used in her construction and 600 horsetails for her rope. She is the largest replica Viking ship ever built, and in her first sea trials reached 10 knots.

The remaining three wrecks were all merchant vessels of varying types; two were designed to carry cargo and it is likely that one was a fishing vessel or a ferry. In comparison with the warships, they were all short, wide and high. The smaller merchantman was built with an open hold amidships, presumably protected by skins when at sea, but was so light that she could have been dragged ashore with ease, even hauled over land for short distances, and could have navigated the shallowest rivers for miles inland. It was the first Viking trading ship ever found.

The larger merchantman was far more substantial in

every way and was clearly never designed to be beached. In comparison to the sharp profile of the warships, this type of trader – known as a *knarr* – had a beautifully rounded shape in both stem and stern, which helps to explain why certain women in Icelandic Viking sagas are described as being *knarr-bringa*, which roughly translates as 'knarr-breasted'. She was built of huge planks of pine, and as pine does not grow to such a size in Denmark she must therefore have been Norwegian in origin. Built for the deep Norwegian fjords, she was heavy but strong, designed to weather the fiercest storms. Her cargo area was also three times the size of the little Danish trader; she had clearly been designed to carry heavy cargoes of goods or people over long distances. Built in this way and to these dimensions, she would have been able to frequent the North Atlantic route, from Norway to Scotland, Iceland, Greenland, and even to North America. Hitherto, no such vessel had ever been discovered; archaeologists knew that the Vikings had spread so far from the evidence of their distinctive tombstones, pottery, clothing and weapons, but never before had they seen one of their ocean-going ships.

Vikings Under Threat

All of the ships had been sunk at the same time, around 1070. Their location blocked off the most direct route to

Roskilde, thus forcing mariners to approach the town by the crooked and winding course, which would have been all but impossible to navigate for those unfamiliar with the fjord. Roskilde itself is one of the oldest towns in Denmark. In the 11th century it was already densely populated, covering at least 50 hectares (123 acres), and surrounded by richly cultivated land. It was also the location of the royal estate of Harold Bluetooth (d. *c.*985) and a religious centre, with four churches, two of which were the first in Denmark to be built of stone.

It is unclear why Roskilde felt so insecure in these years, but it is known that the great Danish successes of the ninth and tenth centuries led to bitter wars between the Norwegians, Swedish and the Danes. In those years, the Danish kingdom was also split between Denmark and England, and domestic enemies may have seized upon the temporary absence of the king to revolt. In 1042, Norway's King Magnus usurped the Danish throne and there was no lasting peace for the next 20 years.

These ships are therefore symbols of great Viking success; they are the varied means by which the Vikings extended their influence across the Atlantic, throughout Europe, even to Seville, deep into Russia, and onward to Constantinople, Baghdad and Jerusalem. All too often, it is easy to overlook the diversity in Viking shipbuilding and to consider them all as generic 'longships', but this collection of vessels reminds us of the range of shipping that underpinned Viking success. In their context as a protective barrier from attack they are also symbolic of the internal strife that ravaged the Scandinavian homelands at the end of the Viking period and which ultimately led to the end of the Viking Age. Danish influence and wealth were severely curtailed when William of Normandy invaded England in 1066, and before long new types of ships and new ways of waging warfare were to relegate these beautiful ships to the indignity of fading memory.

It is a credit to the Danes that the importance of these ships was recognized at the time of their discovery. A new channel through Roskilde fjord was in the process of being dredged a matter of months before the ships were found, and it is fortunate that they survived at all. But it was the sheer commitment, pride and belief in their nation's history, and an uncluttered understanding of the importance of such finds for educating future generations that provided the momentum for these ships to be excavated and then put on display. Roskilde town council quickly offered to find one million crowns to build a new museum and suggested three separate sites. The Danes led the world in seafaring in the 11th century, and they used the remains of wrecked ships to lead the world once more in the preservation and presentation of their maritime heritage.

Galleons and Gold

The growth of cities in medieval Europe stimulated a huge upsurge in trade. After ancient overland routes such as the Silk Road were disrupted by Mongol invasions from c.1220 onwards, seafarers pioneered routes around Africa to India and the Far East to bring the luxury goods of the East to Europe. In the Atlantic, meanwhile, Iberian navigators reached the New World and harvested the riches of South America.

Unfortunately, maritime historians of this period have as little first-hand evidence to work with as scholars investigating the Dark Ages. Although the finds themselves become relatively more spectacular because of the increasing proximity in date to today, they are not significantly more numerous as one might expect. Indeed, in spite of the limitations in our knowledge of Dark Age shipping, we know from maritime archaeology a great deal more about Viking ships than we do about medieval merchantmen, or the ships in which the Portuguese and Spanish discovered the New World and voyaged around the toe of Africa into the Indian Ocean and beyond; very few examples of ships from this period have ever been found. We know almost nothing about the origins of the great Chinese junk that was to dominate Asian waters for more than 2000 years. Between 1000 and 1650, moreover, there was an explosion in different ship types. This was the period of the cog, the cocca, the carrack, the caravel, the nao, the galleon and the galleass. In relative terms, therefore, there is a great deal more for us to learn about shipwrecks of the Middle Ages.

The Growth of Trade

Nevertheless, there were certain highly significant developments that affected the nature of shipwrecks in this period. In general terms, trade increased to a great extent. More ships carried larger cargoes more regularly and further than ever before. The ships became bigger to answer the needs of this trade; although there are instances of large ships or barges from the ancient world, such as the Roman pleasure barges of the emperor Caligula, built in AD 37, which were longer and broader than Nelson's HMS *Victory*, these remain unique examples of such craft from the ancient world. By 1200 it seems to have been normal for ships to be capable of carrying over 100 tons of cargo, and some as much as 250 tons. Around this time too,

Chinese junks began to transport liquid cargoes of up to 50 tons. These are the earliest known ancestors of modern tankers.

The larger ships had more numerous crews to handle the increased quantity of sails and rigging, and those men in turn needed more space for their accommodation and supplies. As long voyages into unknown waters became much more common, these large crews had to be self-sufficient for many months at a time, which, again, increased the need for more space. It is important to remember that these men also needed to return whence they came. Christopher Columbus (1451–1506) took supplies for 15 months on his first voyage to America in 1492, while Vasco da Gama (*c.*1460–1524) took almost 36 months' worth of supplies on his voyage to India from Portugal in 1497. The scale of a shipwreck of this period therefore increased dramatically; larger ships were lost more frequently with larger and more valuable cargoes. The death toll from shipwreck necessarily increased also.

Medieval Men of War

In warfare, the increased size of these ships made ramming an effective tactic, and a collision could easily force a ship to founder, but the increased size of ships had another significant implication. Ships of this period were built bigger by building upwards; they all had a very high freeboard and the military advantages of such height were quickly apparent. This was an age in which battles at sea were fought between men with weapons designed for use on land; the ship in effect acted as a type of siege-platform. Sea battles were more akin to land battles fought at sea than to the punishing exchanges of broadsides that characterize later periods. Indeed, in these years, although cannon were carried to sea, they were never carried in great numbers and they were carried as much for the psychological effect of the noise and the smoke as for the physical effects of their shot. By 1400 heavy siege artillery could be carried to sea, which theoretically could sink an enemy ship, but the likelihood of doing so was small indeed. Ships were more likely to be damaged badly by combustibles hurled manually from one ship to another, and in nearly every instance battles were settled by hand-to-hand fighting.

The advent of cannon, however, had one very significant impact on a vessel's seaworthiness. Bronze cannon were exceptionally heavy, and when combined with their carriages and shot they could easily affect the stability of a ship. They therefore had to be stowed very low down, which required gunports to be cut into the hull. The weight of the cannon also required that they were mounted in pairs, one on each side of the ship. The high fore and aft castles of the ships were crammed with men to make the best use of their bows, cross-bows and other small arms, which made the ships unstable and liable to heel at extreme angles. Gunports cut into the hull less than 1 metre (3ft) above the waterline therefore greatly increased the risk of shipwreck through flooding.

New Seafaring Aids

At the same time, however, the widespread adoption of various navigational aids made seafaring safer. We know that ships in northern Europe were equipped with magnetic compasses in the 12th century, and that by 1410 they were mounted in binnacles to aid with steering. By 1485 they were equipped with gimbals which kept them steady as the ship moved with the swell. Sandglasses were commonly used to measure time, which was another aid to navigation; if it is known how long a ship has been sailing on a particular course, it is easier to estimate one's location. This was further assisted by pocket sundials which allowed a sailor to judge with a good degree of accuracy the time of day. Speed was measured by the log; a wooden drogue attached to a line. The log was cast overboard, and the amount of line that ran out in a given time measured. The speed and time spent travelling on a certain course could then be recorded on a traverse board – a memory-aid for navigating by dead reckoning which worked by placing pegs in a board to record course and speed.

Another great advance came with the introduction of portolan charts, which were marked with 'rhumb' lines. These showed a host of different courses, distinguished by their angle from north. Now, if a mariner knew where he was and where he wanted to end up, he could use a chart to navigate there, sailing on, or parallel with, a rhumb line. The use of these charts was augmented by the quadrant, astrolabe and later the cross staff, which assisted solar and stellar navigation. Although the astrolabe had been a well-known method for measuring the altitude of the sun and stars for centuries, it was not until the mid-15th century that they began to be used regularly by mariners. Indeed, the earliest evidence for the use of the quadrant and astrolabe for marine navigation come from 1456 and 1481 respectively. By the middle of the 15th century, almanacs with tables of the sun's declination, phases of the moon and the location of certain stars became widely available.

All of this equipment appears remarkably impressive when compared to that which was available to earlier mariners, but in every instance it must be remembered that the equipment itself was quite inaccurate, and practical sailors rarely bothered themselves with their exact location until they were within sight of land. Then, using their own knowledge, or perhaps that of an expert local pilot, they could navigate easily and relatively safely; all of this equipment augmented but did not replace traditional methods of navigation. The basic dangers of an unknown shoreline, or a violent storm on a friendly coast, still remained. In that scenario, at least, medieval sailors were slightly more fortunate than their predecessors. Ships were now large enough to be equipped with a stern-hung rudder which made navigation in confined spaces much more controlled than was possible with the cumbersome steering-oar. Moreover, the ships were both bigger and more robust, and from the mid-1300s the largest were equipped with two or three masts. The design of the sails and the placement of the masts now made it much easier to make ground to windward and to manoeuvre; the medieval sailor therefore became less a prisoner of the elements than he had ever been before.

From the mid-13th century maritime authorities in major Mediterranean ports began to introduce legislation concerning freeboard to prevent ships from going to sea over-laden. The first evidence we have of such regulations comes from Venice in 1255, where it became illegal to exceed the draught of a vessel, marked by a cross. We also know that Cagliari, Pisa, Barcelona and Marseilles all had similar provisions around this time. By the 14th century, Genoese statutes were elaborate, with inspection procedures and detailed penalties if the requirements were not met. In the 15th century there is evidence of legal requirements for the seaworthiness of vessels, and the most detailed come from the Hanseatic League, which passed significant statutes concerning safety at sea in 1412, 1417 and again in 1447.

Scattered evidence for lighthouses also comes from this period, although they do not necessarily represent national interest in the safety of mariners, as many were privately owned and run as businesses – ships passing the light to a nearby port were taxed for the privilege of their safe journey. The most famous medieval lighthouse was the great lantern at Genoa, which was tended at one point in the 15th century by the uncle of Christopher Columbus. Other lights shone at Venice and at the gates of the Bosphorus. Important French lighthouses stood at Aigues-Mortes on the south coast and at the mouth of the Gironde, the latter established by the Roman emperor Caligula and rebuilt by Charlemagne in 800. Lighthouses were also located at Dieppe, Harfleur and La Rochelle. Spanish lights shone from the headlands at Corunna and at the entrance to Seville's great river, the Guadalquivir. In Britain, coastal hermitages, convents and monasteries began to build lights and beacons from the early 14th century. A fine example is St Catherine's Tower on the southernmost point of the Isle of Wight, which dates from 1320.

Invasion and Exploration Fleets

These were the years in which maritime enterprise reached new heights. Great fleets put to sea, first with English crusader fleets that voyaged across the Mediterranean, and later with the Spanish Armada's attempted invasion of England. Far in the East, in 1274 the Mongol ruler Kublai Khan (r. 1260–94) launched a vast invasion fleet of 450 ships carrying 30,000 troops from Korea to Japan; it was beaten back by a tempest – the original *kamikaze* ('divine wind') that saved Japan from foreign invasion. Such maritime enterprise carried with it increased risks, and maritime archaeologists today are still unpicking the wrecked remains of both Kublai Khan's fleet and the Spanish Armada.

Great voyages of discovery were also undertaken. In Chinese waters, Admiral Zheng He (1371–1433) undertook seven extraordinary voyages between 1405 and 1433 with a fleet of 62 of the largest junks ever built, with 225 support vessels and 27,780 men. They ranged the Indian Ocean and even explored the east coast of Africa. The ships were an estimated 3000 tons – that is ten times larger than any European ship afloat at that time. They were big enough to transport giraffe, camels and rhinoceroses back to China.

Meanwhile, at the very western edge of Europe, the

Portuguese were beginning to gaze to the western horizon. Their country was poor, and Spain was a powerful neighbour looking over their shoulder. But out to the west, the Atlantic winds blew regularly from the Portuguese coast towards the southwest and to the Canary Islands, Madeira, and from there, deep into the Atlantic. The Portuguese were also equipped with beautiful natural harbours that could easily and safely be located from far out at sea. In short, they had the ways, the means and the desire to explore. Christopher Columbus, a Genoan working for Spain, crossed the Atlantic in 1492, but by 1498 Portuguese explorers had reached India, and by 1500 had discovered Brazil. Nineteen years later the Portuguese Ferdinand Magellan (1480–1521) circumnavigated the world. While these tremendous enterprises were undertaken, trade boomed in European and Asian waters. Northern waters were dominated by the great cogs of the Hanseatic League, and in the Mediterranean by the great powers of Venice and Genoa. Arabic, Chinese, and Southeast Asian shipping criss-crossed the Indian Ocean and China Sea. To protect that trade, powerful navies were born as country after country looked to the sea to increase its wealth and prestige. Most famously, Spain discovered new lands in South America rich with silver, gold and emeralds, and endless supplies of tropical hardwoods with which they could build the ships that they needed to capitalize on their discoveries. As physical barriers were broken by technology, ideological developments followed in suit, emphasizing the benefits of spreading religion and expanding empires, all of which was to be done by sea. As more men took to the sea, more ships found themselves in peril, and more wrecks begin to bear the weight of historical significance. From these years the wrecks are not important or interesting just because they illuminate a certain aspect of trade, warfare or life at sea, but because they can more readily be associated with significant historical events about which we already know a great deal; this is the age in which the history of shipwrecks becomes more readily integrated with the most significant events in the history of the world.

The *Mary Rose*

19 July 1545

The pride of Henry VIII's fleet, the Mary Rose, *did not meet a glorious end in battle, but heeled over and sank as she left her home port. The combination of a tight turn and a refit that had left her top-heavy sealed her fate. She lay largely forgotten for over 400 years, embedded in the deep silt of the Solent. Her salvage from the 1970s onwards was a major event that captured Britain's imagination. The* Mary Rose *has proved a treasure trove for archaeologists, yielding a huge number and variety of everyday artefacts that give a vivid picture of what life was like for Tudor seamen.*

When the *Mary Rose* sank in July 1545, Henry VIII (r. 1509–47) had been king of England for 36 years. By then he had lost one wife to illness, divorced two others and executed two more. He was only to survive for another 18 months. The politics of his reign very much reflected Henry's character; he was ambitious, bold, rich, and he had a claim to the French throne. To realize his ambitions, Henry built a powerful navy that could both transport his troops and fight at sea. Hitherto, these twin requirements of fighting ships were still not mutually exclusive: a powerful ship was one with large numbers of men who would use their small arms to target another enemy ship. In the early 1500s, however, the design of ships began to change, as large guns were mounted on gundecks. Their impact was devastating both materially and psychologically. The *Mary Rose* was one of three great ships that Henry built in England in 1509 to spearhead his new fleet. It is unclear if she was first built with a gundeck to house heavy guns, but by 1536 she had certainly been fitted with one. She was one of the most powerful warships ever to have existed, and bristling with cannon, she heralded a new era of naval warfare that would eventually lead to the broadside gunnery duels of the 18th century.

Three years later Henry began a massive programme of coastal fortification that was unlike anything that had been seen in England since Roman times. This was Henry's response to the annulment of his marriage with Catherine of Aragon (1485–1536) and his resulting excommunication from the Catholic Church in 1533. He now stood alone while the infuriated Pope Clement urged Francis I of France and Charles V of Spain (Catherine's nephew) to unite against him. War with the Catholic Church, moreover, also meant war with Scotland which was ruled by the Catholic James V (r. 1513–42), who maintained strong links to both France and Rome. To make matters worse, in 1544 Henry had further roused the French by capturing the Channel port of Boulogne, and had then granted his subjects the right to unrestricted private warfare at sea. English pirates targeted any neutral shipping with impunity. Henry was surrounded by enemies.

It was in this context that the French assembled a huge invasion fleet in the mouth of the Seine with the aim of attacking Portsmouth, both the premier naval base on the south coast of England and the main source of supplies for the besieged Boulogne, which the French wanted back. The French force consisted of 150 ships and 25 galleys designed to wreak havoc in the confined, calm and shallow waters of the Solent, the narrow stretch of water between the Isle of Wight and the mainland. British naval strength was less than half that, but their defensive position was strong. The narrow entrance to Portsmouth Harbour is still its most distinctive feature today, and modern yachtsmen know that the deep water channel that runs through it is itself flanked by shifting sandbanks in the Solent; these sheltered waters are not as safe as they first appear. That deep channel was guarded by three huge batteries: Southsea Castle, the Square Tower and the Round Tower. Henry took personal command of both the army and the navy and waited in

Portsmouth for the French attack, his command based in an encampment around Southsea Castle.

A Sudden End

The British fleet left the safety of the harbour to face the French threat and a distant and ineffective skirmish ensued until the outnumbered British withdrew, to shelter in the safety of the shore positions. The next day was calm and the French galleys closed in an attempt to draw the British ships from their anchorage. The *Henry Grace à Dieu* was clearly the target for most of these galleys, and just as it was an unwritten rule of sea warfare that much was to be risked to attack the enemy flagship, so too was it that everything must be done to defend her. The alarm was raised and the *Mary Rose* quickly made sail to her support. Her guns were run out, her sails set and her rudder put hard over. But as she turned towards the *Henry*, she started to heel. As she started to heel so water began to lap at the lowest gun-ports. Then it began to pour in.

As the heel of the ship worsened, guns and cargo that were lashed to the deck or bulkheads suddenly found gravity pulling them in an entirely unexpected direction. Cannon fell from their mountings and rolled to the lowest point of the deck, while shot, cargo, men and belongings all shifted to starboard. Soon the angle of heel was such that the ballast shifted. By now, just minutes after setting sail, the flow of water into her hull was irresistible and she was forced underwater at startling speed. With her hull no longer buoyant, that weight, combined with the 600 men, their stores, cargo, ballast and guns simply drove the *Mary Rose* towards the seabed with such force that her keel came to rest 3 metres (10 ft) below the estuarine mud of the Solent floor. Every report of her loss agrees she sank so quickly that only a handful of her crew escaped. The French were unable to take advantage of British shock at the loss of the *Mary Rose*, and they withdrew out of range of the British guns to the Isle of Wight, where they launched a minor attack with little effect.

That, then, is the story of the loss of the *Mary Rose*. It is not in any way glamorous or overly dramatic, nor is it soaked in the terrifying power of the sublime: there was no tornado, no tsunami, in fact there was no wind or rain at all. It was, rather, a pleasant British summer's day. One contemporary French chronicler quickly made the leap between the sinking of the *Mary Rose* and the presence of French warships to claim that she was sunk by French gunfire, but we know that to be untrue. She sank almost entirely unprovoked. As such, the loss of the *Mary Rose* is often presented as an isolated tragedy, but in this period the whole business of making war at sea was unpredictable and highly dangerous. For example, just a week before the *Mary Rose* was lost, her equivalent in the French fleet, the flagship of Claude d'Annebault (1495–1552), the *Philippe*, had accidentally caught fire and been abandoned in favour of the *Grande Maîtresse*, which promptly ran aground.

Yet the *Mary Rose* sank as the result of various repairs and improvements, the most significant of which was the introduction of a gun deck in 1536: the *Mary Rose* of 1545 barely resembled the ship launched in 1509. She carried more men and guns than originally envisaged and the top-heavy design of medieval carracks left little room for error concerning stability. Thus, in 1545 her design was fatally flawed, leaving her prone to sink at any time had she not gone down that day. With some degree of fortune, however, she did sink in that location and in the way that she did. She landed on her starboard side, the sturdy hull protecting almost all of her interior from the powerful impact of her fall. The tons of soft upper sediments that she disturbed quickly buried her again in a layer of silt that protected the submerged half of her hull and everything inside it from the passage of time. There, 2 metres (6 ft) under the seabed there was no daylight to fade colours, and no animals to destroy timber and organic material.

As with all wrecks of value or historical significance that occurred in relatively shallow water and within sight of land (as most wrecks in fact do – entering and leaving port being the two most hazardous parts of any voyage), they are usually subject to disintegration on the seabed or to salvage attempts. The *Mary Rose* is no exception.

Immediately after her sinking, 30 Venetian mariners tried to salvage her under orders from the king, and it is from this attempt that the ship's mainmast was torn from its footing. Documents also survive detailing rewards for guns and other valuable items recovered from the wreck in the late 1540s, but after 1549 there is no evidence that the wreck was interfered with until 1836 when it was rediscovered by John and Charles Deane, the entrepreneurial brothers who invented the diving suit.

Thus the *Mary Rose* was subject to some degree of human interference after her sinking, and over time the currents of the Solent and the animals that live there wore away at the exposed hull and its contents until the entire port side was so weakened that it was simply washed away with the tide. What was left was the starboard half of the hull and that is what you can see on display at the Royal Naval Museum in Portsmouth where the hull stands under a twilit shower of polyethylene glycol, a chemical that will permanently stop the process of disintegration.

Tudor Time Capsule

The means by which the hull got to be there is a saga in itself. She was found in May 1971 after six years of searching, her timbers protruding from the seabed like the fingers of a body buried in the silt. By 29 June the excavators had found their first major Tudor artefact, an iron muzzle-loading gun. Gradually, the hull also began to show itself. The oak shone a warm brownish red even in the polluted waters of the Solent; the treenails that fastened the planks to the frames were tight and sound; at least 9 metres (29 ft) of the hull, measured from the keel, described the unmistakable shape of a medieval carrack. This, in itself, was astonishing, but still more surprises were in store, for the archaeologists soon came to realize that the hull cradled a time capsule of Tudor seafaring.

There are far too many stunning finds to name piece by piece, but suffice to say that most of a leather jerkin was found, intact, with a comb in its pocket; a bosun's call was found, still attached to its original ribbon; at least 27 leather ankle boots of which 10 are definite pairs were found, but these only comprised 17 percent of all the leather footwear recovered. All in all there were 655 items of clothing, from silk hats to woolly socks. Numerous prayer books and rosaries were discovered, plus inkpots, pens and dice. The barber surgeon's chest contained 11 wooden ointment pots, each with residues of their original ointments, along with his shaving kit and a wooden feeding bottle for those severely sick or with facial injuries. There were drums, pipes and fiddles, one with its own wooden case. The door of the carpenter's cabin was wedged slightly open, leaving just enough room for a man to squeeze through. The cabin was piled with chests, full of work tools. Many of his planes had holes drilled through the heel end; presumably to hang them on the wall when ashore (hanging tools on the bulkhead of a pitching and rolling ship would be a bad idea).

In the hold, at the very base of the ship were two huge brick-built ovens with copper-alloy cauldrons, surrounded by casks, chests and prepared logs for fuel. Nearby was a basket of plums and another of fish. More than 60 serving bowls were found, half of which were marked with initials or knife cuts, presumably to indicate ownership, or perhaps the manufacturer. There were 179 skulls, perhaps 43 percent of the entire crew, 68 with matching jaws. Two of those skulls were children's, probably around ten years old. One skull had a healed arrow wound while three had broken noses. Regular and compulsory use of the longbow had distorted many of the skeletons, and there were numerous examples of osteoarthritis.

Because of the remarkable state of preservation, what was not found was in many cases equally instructive. There was, for example, only one surviving chamber pot, indicating that the men simply relieved themselves over the bows, where the waves would keep the ship relatively clean. In 1545 hammocks were not used by sailors but a few examples of thick wool survived, suggesting the use of blankets. Nor was there any evidence of tables or seating; the men appear to have sat on deck to eat their food, play games, read books or sleep. The small amount of money found suggests it is unlikely that any games were played for cash stakes.

It is clear that when the ship heeled, objects were hurled with astonishing force against bulkheads where they smashed, or even broke through. One of the great cauldrons from the galley burst through to the deck above. Six men were found together near a bronze culverin; drowned at their station as they tried to escape. Many bodies were also found at the foot of the ladder providing access in and out of the galley; a bottleneck of desperate men. From the depths of the ship, however, there were very few rat skeletons: long before the crew realized the gravity of their situation, the rats knew that the ship was doomed.

La Trinidad Valencera

26 September 1588

Harried by the English fleet and scattered by bad weather, the Spanish Armada of 1588 was forced to abandon its attempt to invade England. One Spanish ship that tried to run for home by rounding the north of Scotland and sailing down the west coast of Ireland foundered off Donegal. Almost four centuries after her loss, La Trinidad Valencera *has provided historians with fascinating new insights into 16th-century military hardware and tactics and the Armada campaign itself.*

In mid-September 1588, the local inhabitants of the north coast of Donegal saw a lumbering Spanish warship slowly approaching land, apparently making for the shelter of Lough Foyle. Before she got there, however, she struck a submerged reef at the western end of Kinnagoe Bay and stuck fast. As the word spread, many more gathered on the cliffs and the beach to watch these strangers wade ashore from their tiny ship's boat. By the time the first Spaniards had made land, finely clothed and carrying as much of their expensive personal belongings as they could with them, there was perhaps a crowd of 20 or so who met them on the beach. The Spaniards were plainly exhausted. An experienced eye might have noted that the ship was very low in the water, and may even have seen in her attempt to run for a foreign shore the desperate move of the captain of a sinking ship. A less experienced eye would have noted that the ship was large, many of the crew ill or injured, and those who had made it ashore were unmistakably rich. The Irish response was unequivocal. They were prepared to help the Spaniards; they would even lend them one of their boats to help more survivors make it to land and to salvage the ship's cargo, but anything that was brought ashore was theirs. They started by relieving the initial landing party of everything that they had brought ashore with them '*to the value of 7300 ducats*'.

The story by which the ship came to be there is part of one of the most remarkable tales in world history. It is the story of Philip II of Spain's failed attempt to invade England in July 1588 – the Spanish Armada.

On Collision Course

In the last quarter of the 16th century, the Catholic Spanish empire was formidable and still expanding. The riches that had started to pour in from South America helped to fund the expansion of Spanish interests at home. Portugal was occupied in 1580 and the Azores conquered in 1582. Soon after, the Netherlands were attacked by Spanish forces in 1583 with great success, giving the Spanish direct access to the North Sea. The Spanish had also made great territorial advances in Italy and even in parts of the Far East. In these years Philip was considered powerful enough '*to wage war on all the world united*'. Queen Elizabeth I of England (r. 1558–1603) would have to fight for her crown, and in 1586 she unleashed her most aggressive admiral, Sir Francis Drake (*c.*1540–95) on the wealth of the Spanish West Indies. The expedition was designed both to raise money for a forthcoming war and also to target the foundations of Spanish might. Drake captured Santo Domingo, the oldest Spanish city in the New World. In less than 24 hours, the magnificent port of Cartagena (now in Colombia) followed. Philip's response was to prepare an army and a fleet for the invasion of Protestant England.

The events that followed are well known: setting sail from Lisbon, after a brief stop at Corunna to replenish their rotten victuals and leaky casks, the ships of the Spanish Armada entered the Channel and were sighted by the English fleet on 20 July. Their goal was to achieve temporary naval supremacy in the Channel, which would allow the duke of Parma's army, waiting in Dunkirk, to make the crossing to England. The Armada would then land its supplies of food, water, munitions and support troops while Parma marched on London.

A full week of skirmishes followed, but only one Spanish warship, the *Nuestra Señora del Rosario*, was captured, after her rigging was damaged. Drake launched a particularly fierce attack as the Spanish neared the Solent, one of the few places in the Channel where the Armada might anchor safely. The duke of Medina Sidonia (1550–1615), the Spanish commander, was forced to anchor instead off Calais. The English response was swift; on the night of 28 July they launched a fireship attack, which drove the Spanish from their anchorage in disarray and allowed the English fleet to close. English gunnery was markedly superior and in the ensuing Battle of Gravelines many Spanish ships were badly damaged. Their gunnery may have been poor, but Spanish seamanship was good – the result of so many successful trans-Atlantic bullion convoys – and they gradually restored a compact fighting formation that the English could not break. The weather soon worsened, however, and the horizon to the east and the north filled with blue: the Armada had been blown into the North Sea.

A Fateful Decision

Medina Sidonia might well have anchored in a safe port to renew his attempt to reach Parma and launch the invasion, but instead, and fatefully, he decided to sail back to Spain around the top of Scotland. Food stocks were low, his ships badly damaged and many of the crews injured or unwell. But Medina Sidonia was as resolute a mariner as he was skilful, and successfully guided back at least two-thirds of his fleet to Spain. Even so, in spite of his explicit warnings to his captains to avoid the coast of Ireland, as many as 35 ships failed to return. Thus far, the wrecks of 14 have been found scattered around the Irish coast. One of those unlucky ships, and one of the largest in the entire fleet, was *La Trinidad Valencera*. Heavily damaged in the Channel fighting, she had just taken aboard 250 survivors from the *Barca de Amburg* when she struck the Donegal shore.

She did not break up immediately but remained perched on her rock for two days until the weight of the hull, now bearing down with full force onto her keel, broke her back and split her in two. She sank with a number of Spanish sailors and Irish looters aboard. The ship's captain, Don Alonso de Luzón, led the survivors to the local seat of power, Illagh Castle, home to Sir John O'Doherty. A powerful garrison of English troops from the nearby Burt Castle heard the news and descended on Illagh. In no position to fight, the Spaniards surrendered. Those who could be ransomed were forced to march over 100 miles (160 km) to Drogheda where they were held. The common sailors and soldiers who had survived the battles with Drake and braved the storms of the north coast of Scotland, and some of whom had been wrecked twice, were butchered in a nearby field.

It was not until 1971 that the wreck was found by divers from the City of Kerry Sub-Aqua Club. Their first discovery was a 3-metre (10-ft) long bronze gun. Nearby they found a large wooden spoked wheel and an even larger gun, engraved PHILIPPVS REX 1556, clearly visible underwater. Better was still to come. Almost 400 years had passed, but much like the *Mary Rose* (see pages 83–6), part of *La Trinidad Valencera*'s hull had embedded itself in the seafloor; the regular tidal surges around the wreck had dug a perfectly fitting grave.

The *Valencera* wreck is symbolic of the entire Armada campaign in several ways. Firstly, she is not a Spanish warship, but a merchantman from Venice, seized in Sicily, and then converted from a grain ship to an armed transport in Lisbon. Her wreck yielded numerous fragments of Italian pottery, while Chinese Ming porcelain testified to her trading past. She formed part of the Levant squadron, which consisted of ten similar converted merchantmen, hailing from Ragusa, Venice, Sicily and Naples, but none were as well armed as the *Valencera*. They all carried a large number of soldiers in their spacious decks, who would lead ship-to-ship fighting against the English and secure the beachhead once a landing had been made. The *Valencera* alone carried 281 soldiers.

To assist with the hand-to-hand fighting, in addition to their traditional armament of swords and heavy muskets, these soldiers were equipped with two different types of incendiary weapons, *alcancías* and *bombas*, both of which were recovered on the wreck site, one example of an *alcancía* still with residue of carbon and sulphur on its inside. The *alcancía* was essentially a ceramic pot filled with gunpowder and thrown by hand. The *bomba* was a wooden tube from which an offensive rocket or firework could be fired. They were stored in wooden barrels without iron bindings to reduce the risk of sparks, but once lit, the danger of using these weapons on a tightly packed wooden ship does not bear thinking about, and in his instructions to the troops before they left Lisbon, the duke of Medina Sidonia instructed that these '*should only be entrusted to the most experienced men* …'. It is an order that at once brings to mind an unruly rabble of young soldiers, but perhaps more tellingly it reflects the character of Medina Sidonia, a compulsive worrier and pessimist, whose gloom was fuelled by his considerable experience of warfare and amphibious campaigns; he was, above all, a realist.

It is especially significant that any such incendiary weapons survived. They were designed for close combat, usually for boarding actions, and the *Valencera* was not issued with a vast number – perhaps as few as 25 *alcancías* and 15 *bombas*, but there was never any opportunity to use them in action. Most of the English ships were far more mobile than these sluggish merchantmen laden with soldiers, and they were careful not to approach too close to the numerous, skilled and well-armed Spaniards, but rather stood off to make use of their great guns. Indeed, had they closed, the outcome might have been far different. Interestingly, such incendiary weapons had been long abandoned by the English, and their concentration on mobility and firepower was to revolutionize naval warfare and catapult them into the era of broadside gunnery that dominated the forthcoming great Age of Sail.

Impressive Firepower

The *Valencera* also carried three huge 40-pounder cannons, each weighing 2½ tons. Each one is a beautiful object, inscribed with the name of Phillip II, his wife, and a list of the dominions that he ruled over in 1588. They are also inscribed with the year of manufacture, the exact weight of each gun and the names of the Flemish gunfounder who cast them and the Spanish official in charge of the siege train. They epitomize 16th-century Spain itself; ornate, expensive, violent and proud. At first, it was assumed that these huge guns were indicative of new Spanish fighting tactics; these were heavy guns to be used aboard ship to target the enemy vessels, rather than to close and engage in hand-to-hand combat, the traditional tactics of war at sea in the Middle Ages. But it soon became clear that these guns had been stowed with associated land-carriages, two full sets for each gun. The wheels were beautifully turned elm heartwood, with oak spokes to carry the great weight of the guns, and their rims bound in iron. These were not guns to be used aboard ship, but for warfare on land. They were not armament, but cargo; they were part of the supplies for the invasion army once it had landed.

The total siege train was to consist of 48 such *cañones*, and 12 smaller 25-pounder cannon known as *culebrinas*. Wielded by 55,000 troops, this was enough firepower, Philip reasoned, to bring London to its knees. To assist the siege further, the *Valencera* carried tripod hoists to mount the guns on their carriages and screw-jacks to change the wheels if they broke. There were planks and great baulks of timber to build gun platforms, *gabions* (basketwork sandbags) and matting to support defensive earthworks, and great bundles of young fir trees trimmed to a sharp point. This was the 'barbed wire' that protected the trenches of their day. There were also rollers and hammers, stakes and axes, shovels, lanterns and buckets. There was even a tent, complete with pegs.

The gun carriages that were found for use at sea were markedly different from those that the English used, and a modern reconstruction confirmed contemporary reports that the Spanish gunnery was incredibly slow in comparison to the English, taking at least twice as long to load and fire. Gunnery equipment from the wreck was also found to be of poor quality, which no doubt had a significant effect on the gunnery itself; gunners' rules and shot gauges used to measure the bore of a cannon and select an appropriate shot were badly inaccurate, with no basis in rational mathematics. For an individual ship, this need not be a problem, but another gunner's rule was soon discovered on a different Armada wreck and it revealed equally large errors, but of an entirely different kind: Spanish gunners, it would seem, were held firmly in the grip of a conspicuous muddle.

La Trinidad Valencera sums up the Armada in microcosm. A bold enterprise that the Spanish alone could contemplate in the 16th century, it was let down by logistical failures and a lack of realism that began with the king himself. The plan for the invasion was underpinned as much by faith as it was by logistics. Just before they left Lisbon, one senior officer wrote: '*we are sailing against England in the confident hope of a miracle*'. The loss of the *Valencera* and her sisters around the Irish coast was the shipwreck of that hope.

The *Vasa*

10 August 1628

The Vasa *was built as a floating demonstration of Sweden's economic and military might during its heyday in the 1600s. But on her maiden voyage in 1628, just yards from shore, she suddenly sank, taking half of those on board down with her. Salvage attempts failed, and the cold, fresh waters of the Baltic held the* Vasa *in suspended animation until, in the early 1960s, a unique prize was lifted from the sea – the world's only intact warship from the 17th century and one of the best-preserved such ships of any period.*

The Baltic is shallow and mainly made up of fresh water, fed by three substantial rivers, the Vistula, Dvina and the Oder. These connect the Baltic with a great hinterland that penetrates deep into the continent and connects with even greater rivers such as the Rhine, Weser and Elbe. The shallowness and freshness of the Baltic is often enough for cursory observers to consider it little more than a lake, but it is a sea in its own right, connected to the North Sea and thence the Atlantic via narrow channels between modern Denmark and Sweden. It is also vast. Although no point in the Baltic is further than 67 miles (107 km) from land, it extends almost 1000 miles (1600 km) from the extreme west to the east. The Baltic has wonderful natural harbours that could easily be exploited to realize the wealth of the surrounding natural resources. In the 17th century it supplied most of Europe's shipbuilding timber, in particular those beautiful tall firs and pines that had the inherent flexibility ideal for the masts and spars of sailing ships. There was also tar and pitch, used to weatherproof hulls and rigging, plus flax and hemp for sails and rope. These commodities were supplemented by substantial exports of grain, and a near monopoly of copper and iron-ore.

In the 17th century the demand for these items, all crucial for waging war and for trade, grew steadily as expanding continental states flexed their muscles. In turn, towns and cities exporting these goods out of the Baltic generated their own demand for British and Flemish textiles, French and German wines and all of the luxuries from the East and the Americas. The commercial and military significance of the Baltic therefore increased, and it became the focus of a power struggle between the numerous states that lined its shores. In the 17th century Denmark, Norway, Sweden, Prussia, Russia, Poland,

Brandenburg and the Netherlands all had a significant interest in the fate of the Baltic.

Sweden's Meteoric Rise

Of those powers, it was Sweden that rose the furthest and the fastest. In the early 16th century, a strong Swedish state formed under a young and dynamic nobleman named Gustavus Vasa (r. 1523–60), who was crowned Gustavus I in 1523. To protect Sweden's boundaries the Vasa dynasty set about establishing formidable armed forces, and because so much of Sweden is coastal, this required a significant navy. The national tradition of Swedish merchant shipbuilding did not, however, lend itself to warship construction. Warships had to be built from scratch or merchantmen captured from other countries and then converted. The pool of sailors that populated the Swedish merchant marine was also insufficient in both numbers and training to man a navy, and so the manpower had to be imported along with the expertise to build the ships. Yet Sweden had no trouble attracting this manpower, and immigrants flooded in, principally from the Netherlands. With the Vasas at the helm, Sweden was clearly a stable and ascendant country, whose reputation relied on the potential of her formidable natural resources, which the Vasa dynasty began to realize through effective state control. Her neighbours, by contrast, although equally endowed with natural resources, were divided and weak and unable to mimic the Swedish model.

By 1630 the Vasa dynasty had time and again demonstrated the value of sea power not only to break blockade and transport troops, but primarily to secure trade, particularly through the easily defensible narrows between Sweden and Denmark that gave access to the North Sea. Sweden had also become a great European military power. Now ruled by King Gustavus Adolphus II (r. 1611–32), Sweden dominated the Baltic, much of Germany and Poland and played a crucial role in the Thirty Years' War (1618–48). Indeed, until the start of the 17th century the Swedish navy was far larger than both the Dutch and Spanish navies, and from 1600 to 1650 rivalled and frequently outnumbered that of the British. Naval stations were maintained at strategic points throughout the Baltic, at Kalmar and Stockholm, Åbo and Viborg, Kexholm, Riga and Elbing. Swedish warships could be found everywhere from Wismar to Reval (Tallinn), Barösund to the Vistula. Swedish success on land was also impressive.

It was in these years of developing Swedish might that the warship *Vasa* was built. Ordered by Gustavus II in 1625, she was designed by a Dutch naval architect working for the Swedish crown, and built at the royal dockyard in Stockholm, opposite the royal palace. When she was completed, she was, in terms of firepower, the most powerful warship in northern Europe. Christian IV of

Denmark and Charles I of England had bigger ships with more guns, but none had the firepower to match the *Vasa*. The guns themselves, moreover, were highly significant. Hitherto a ship's armament had been formed of a variety of gun types, sizes and designs. A ship might expect to be armed with a mixture of cannon, demi-cannon, cartowers, sakers or minions, and every one of a different size, design and construction. Now, for the first time aboard the *Vasa*, they were of a standard calibre. She was armed with 48 24-pounder cannon.

A Traumatic Loss

On 10 August, 1628, she left the safety of the royal dockyard manned with 100 crewmen plus their wives and children, and amidst a carnival atmosphere she made for open water and Älvsnabben, where she was to be kept in reserve, and used to augment the Swedish fleets off the coasts of Denmark or Prussia as required. Before she had gone half a mile, however, she began to heel over to port. Frantically, the guns on the port side were hauled to the starboard in an attempt to right the ship, but to little avail. The ports were all open and the lowest very close to the waterline. The sea soon breached the hull and, like a river breaking its banks, poured into every opening it could find. Before the eyes of thousands, this stunning ship, which moments before had held everyone's attention as her sails proudly filled and her pennants and flags danced to the whistling of the Baltic wind, simply disappeared.

We know from diaries of witnesses of more recent events that such a tragedy can have a profound impact on onlookers, which should not be underestimated. It is a sleight of hand the magnitude of which is hard to conceive. When filled with water a ship can disappear suddenly and violently, and the waters close shut as if nothing had happened. The creaking and smashing of timbers and the screaming of those dying suddenly stop, and are replaced by a pregnant silence. To the sailor on board, moreover, the emergency is realized far in advance of a distant observer who seldom conceives there is something wrong until it is too late. The celebratory mood on the day made the loss of the *Vasa* all the more poignant; the impact of the disaster must have been strikingly similar to the loss of the space shuttle *Challenger* in 1986, when she exploded on lift-off.

There is no real evidence of what happened in the immediate aftermath of the sinking, although presumably rescue operations were mounted for the survivors. But as with most large ships and all important ones, there were immediate attempts to salvage her, and in this instance the job fell to an Englishman named Ian Bulmer. He was unable to refloat the ship but he did manage to get her back onto an even keel, which made her salvage in the 20th century significantly easier. His was the first of a series of attempts, but none worked. The ship stayed where she was, perched on her keel on the bottom of Stockholm harbour while she was stripped of her bronze cannon by salvors. By 1665 most of her armament had been recovered.

Gustavus's Floating Palace

Otherwise she remained untouched. When she was rediscovered in 1959 she had not been damaged by enemy fire, by storm, animal or plant life, by ice, heavy currents or tides: the heavy clay and fresh, cold and calm waters of Stockholm harbour had simply covered her in a protective balm. The quality of her survival was not unlike the *Mary Rose* (see pages 83–6), but with the *Mary Rose* only a third of her hull survived the salty and swirling waters of the Solent. In contrast, the entire hull of the *Vasa* was intact. Almost 2830 cubic metres (100,000 cu ft) of 17th-century life had been preserved at a precise moment in time. Everything on board, moreover, was brand new. The value of her contents for the archaeologist was therefore exceptional. It was possible to say that every item on board was in use at the same time. This is not a luxury that many archaeologists can enjoy, since the artefacts in most sites reflect many years of inhabitation, as heirlooms are passed down from generation to generation, or other long-forgotten items are placed in storage.

Divers first excavated tunnels under her hull through which they threaded steel wires. These then cradled the ship as she was pulled free from her coffin of clay. She was gingerly raised until the top timbers just protruded from the water and then she was pumped out. Such was the quality of the hull's preservation that it only required a few basic repairs to ensure that the *Vasa*'s hull was watertight. Slowly she emerged from the waters, and floated on her own keel into a dry dock. It remains the most remarkable salvage in history.

Her decks were tightly packed with clay which had to be painstakingly excavated, but the finds were stunning. Everything was preserved in fine detail, even bodies of the crew and their wives. One body was complete with hair, fingernails and shoes. Many of the carved decorations had fallen off but by carefully matching nail holes and joins, most were replaced, and she now stands 95 percent complete. The museum housing her is also an exquisite creation, where the ship can now be seen up close, at six different levels, as well as at a distance.

The carvings epitomize the art of the Renaissance and northern European culture. The ship's figurehead is a huge lion (the Vasa family emblem). Three metres (10 ft) long, it snarls its defiance at the sea. There are over 700 different carvings, including Roman emperors, figures from the Bible, classical mythology and Gothic heroes. Recent research has shown that the background colour was red, the carvings bright and multi-coloured. Some gilding accentuated certain features such as hair and beards. The *Vasa* truly was a floating palace.

Some 16,000 artefacts were recovered from her, illuminating every aspect of life at sea, but the one unresolved question is the cause of her sinking. There is very little information other than the wreck to work from: there were no contemporary enquiries and the *Vasa*'s plans had been personally approved by the king. Her builder Henrik Hybertsson also died before she was launched. Nevertheless, we are gradually beginning to piece the facts together from the evidence of the ship herself.

A Long-running Mystery

Numerous theories have circulated about the *Vasa*, most notably that her design was altered by the king, who insisted that she be made longer and with an extra gundeck. It certainly is a good story, and in Sweden, as in so many constitutional monarchies, it is politically entertaining to blame the king. But the *Vasa*'s proportions are perfectly natural, unlike those of a ship whose design was altered at a late stage. In fact she is the very model of a 2-decked warship of the 1620s. Certainly, the guns of her lowest deck are perilously close to the waterline, yet this was a common feature of all ships of that era; no one had yet discovered how to raise the decks to a safe level while maintaining an acceptable level of stability. In fact, the *Vasa* 's lowest gunport when she sailed was 150 centimetres (5 ft) above the water – the same height as the lowest gundeck of HMS *Victory* when she was built in 1765. And *Victory*, it must be remembered, was built as an ocean-going warship designed to weather the stormy Western Approaches, whereas the inland waters where the *Vasa* sank were mirror-calm.

The height of the gundecks cannot, therefore, explain her sinking. There is also no structural evidence that she was altered at any stage, other than a slight widening of the ship by 43 centimetres (17 in) which would, unquestionably, have *improved* her stability. There is written evidence of disagreement between the king and designer, but it was over before the keel was laid. Contrary to popular myth, then, all the evidence suggests that the *Vasa* was built exactly as she was designed.

We do know that she was a very tender ship, however, which means that she rolled heavily and easily. In her stability test carried out just before she sailed, 30 sailors ran from side to side to generate a rolling motion. This test was soon stopped because she rolled so much. Yet there are many ways to temporarily control heavy roll, through clever ballasting, manipulation of the sail plan and choice of course and speed. Above all, a crank ship not ballasted correctly should keep her gunports closed. But the *Vasa* sailed with them open and her hull incorrectly ballasted.

Moreover, her minimal crew of only 100 further compromised her stability. With more men would have come more stores and more weight, and although this would have brought her lowest gunports even closer to the water, she would have been less likely to heel suddenly. The crew was also too small to respond quickly to an emergency. Thus, with the benefit of hindsight, the tragedy of the *Vasa* is shocking, but certainly not surprising.

She is now preserved in the Vasa Museum in Stockholm, where she is visited by more than 900,000 people each year (and a total of more than 25 million to date). The *Vasa* was built to be a symbol of Swedish prestige and power in the 1620s, after 333 years at the bottom of the sea, that is exactly what she has become.

Broadsides, Pirates and Whales

By the end of the 17th century, most of the major design problems facing shipwrights had been solved. The use of ship plans was standard and the mathematics of extrapolating the dimensions of a full-sized ship from a model was better understood. Hulls were still relatively short in comparison with the lengths that they would reach by the 1820s, and the biggest ships were also relatively crank – they would roll heavily in a seaway and were in danger of flooding through their lowest gunports. Nevertheless, the vital ingredients of the great warships of the classic Age of Sail had been achieved; these ships carried large numbers of guns on one, two, or three decks.

The hulls of these ships were still made entirely of wood, and as a result they tended to rot over time. Tropical hardwoods like teak were less prone to this, as they are almost impervious to rot, but they could still suffer from attack by the shipworm, *Teredo navalis*, which burrowed into the timbers. Another problem with all ships' hulls is the growth of weed and barnacles that reduces both speed and manoeuvrability. To reduce the impact both of the shipworm and fouling, the Royal Navy began experimenting with nailing copper onto ships' bottoms from the 1750s. At first, however, the electrolytic action generated between the iron fastenings in the hull and the copper sheathing was not understood, and the hulls gradually began to fall apart. Until this problem was solved, coppered ships were far more vulnerable to wreck than they had been before. The *Royal George*, Hawke's flagship at the famous Battle of Quiberon Bay in 1759 (see pages 136–42), foundered in the calm waters of Spithead in 1782. It was believed by many that her copper was to blame.

For the first time, wars were now fought between European nations in distant waters; in North America, the Caribbean and the Indian Ocean. While knowledge of those waters grew as trade and its protection came to them more regularly, repairing damaged ships abroad became a major headache. Ships cannot be repaired with any old timber; it must be seasoned and of good quality. If it is not, then it will rot quickly, or if used in the rig it will be unable to withstand the forces required of it and it will crack. Ideally, the ships would also be dry-docked to facilitate repair to the hull, but building and maintaining such facilities thousands of miles from home was a formidable challenge and it was not always met. Even with Spanish,

British, French, Dutch and Portuguese colonies in and around the North and South American coasts and the southern coast of India, to sail in waters so distant from substantial European naval infrastructure was still very much a risky business, and many ships that were damaged there never returned to Europe.

The Curse of Scurvy

The ships themselves were not the only concern in foreign waters. Disease spread rapidly among northern European sailors, unaccustomed to the damp heat of the tropics, and genetically unprepared for the biological assault that was inevitable from a prolonged voyage in those latitudes. Scurvy, caused by a lack of vitamin C in the sailors' diet, could weaken a crew as effectively as any disease. A ship with a weak crew was a weak ship, and far more susceptible to shipwreck.

It was understood among seafarers from as early as 1600 that a diet of fresh fruit and vegetables could both prevent and cure scurvy, but the ability to acquire such victuals was always a problem. It was not, for example, only a matter of acquiring lemons for a handful of men; the largest warships of this period carried crews approaching 1000 men. Fleets of 20 or more ships were not uncommon. These fleets shared a population size with a substantial town of the 18th century and the administrative and logistical infrastructure required to keep them safely at sea was a challenge that was difficult to meet in home waters, and always a problem on foreign stations. In spite of the improvements in the ships, therefore, the men who sailed them still remained vulnerable.

Technological Advances

After 1762, however, those sailors were equipped with a piece of technology that radically improved their ability to navigate with safety. Hitherto, sailors out of sight of a known stretch of coastline had never known with any degree of accuracy exactly where they were. The source of the problem was that it was impossible to establish one's longitude at sea. Latitude could now easily and reliably be ascertained by observation of the sun through a sextant, but to calculate longitude required an accurate reading of the time of day. This, in turn, required the invention of a clock that would not significantly lose its time aboard a rolling, pitching, wet, hot, dry or damp ship for months, if not years at a time. It was a problem that had bemused scientists for many years, but it was solved by John Harrison (1693–1776) in 1761, when he invented his chronometer, the H-4.

Nevertheless, to be comforted by knowledge of where you were first required a certain degree of knowledge of what there was near to you. The impact of safety at sea from the introduction of the chronometer, therefore, is intricately linked with the production of accurate charts which required skilled hydrographers to be sent on sponsored voyages to map both the known and the unknown world. They also needed the requisite skills and these were not available until it was realized that if a quadrant or sextant was turned sideways it was possible to measure horizontal angles. Then, by measuring two or more angles to known points, one's location could be pinpointed; a process known as triangulation. The first accurate chart of the British Isles was begun in this way only in 1744, but hydrographers quickly spread their gaze all over the world and within a century very little of the world's coastline remained unmapped.

If one was unlucky enough to suffer a shipwreck in this age, there were still few technological advances to be of any assistance, apart from the introduction of the davit. All large ships of this period carried at least one, and in some cases several small boats, usually stowed amidships between the fore and main mast, above the hatches. They were then launched over the side of the ship using blocks and tackles attached to the lowest yards on the fore and main mast. The boat would be raised, swung over the side of the ship and then lowered. Although it was possible to prepare the blocks and tackles ready for an emergency situation, launching a boat was still a relatively time-consuming and labour intensive operation. Davits made it much easier after their introduction in the 1790s. These were hook-shaped supports that suspended a ship's boat over the side or the stern. The boat would then only have to be lowered by a small number of crew. It saved valuable time, and in so doing saved many lives; it was now far less likely that a large ship would go down with her boats still aboard.

Lifesaving Innovations

Yet even if this did happen, there was still some hope for shipwrecked mariners if they were fortunate enough to hit a stretch of coastline that was now equipped with a coastguard. History is littered with accounts of fishermen taking their boats out to save wrecked sailors, but not until the late 18th century did any permanent force exist that was responsible for caring for those who had been wrecked. Even so, when the British coastguard was finally established at the end of the 18th century, this was not its primary function, but subsidiary to its role in preventing smuggling. When coastguards attended a wreck, their principal responsibility was to protect the nation's import duties. They were first to secure the wrecked cargo from the local population, who would quickly strip any wreck of anything and everything of use or value.

Lifesaving was a subsidiary responsibility, but it gradually grew in importance. Coastguard stations were soon equipped with Manby's Mortar, a device that had been tested and officially approved by the Ordnance Board of the Admiralty. Invented by Captain George W. Manby (1765–1854), a boyhood friend of Nelson, it fired a line from ship to shore which could then be used to winch people and cargo to safety. This was critical, as all too often a ship wrecked on a rocky shore is impossible to approach from another vessel. Manby had personal experience of the frustration that this could cause, having helplessly witnessed 67 men die aboard the brig *Snipe* which was wrecked off Great Yarmouth in 1807. Although only 100 yards out, no one was able to reach her. His heaviest mortar could now fire a line up to 500 yards. Every station was equipped with one, and every member of the crew instructed in its use. Manby's invention led to a host of similar devices of varying quality that adapted mortars, and then rockets, to this use. Although the earliest figures for the coastguard are from 1856, we know that in that year alone 2231 lives were saved from shipwreck by the coastguard, and by the time of Manby's death in 1854 over 1000 lives had been saved by his mortar alone.

Lifeboats also had become common around the British coasts by the end of the 18th century, and in Liverpool there was even a system of awards in place for the use of the local lifeboat. At Bamburgh near Newcastle, an elaborate system was set up. In stormy weather two men patrolled the shore. If a ship was spotted in difficulty one reported to the nearby castle to raise the alarm, and a gun fired from the castle's tower was a signal for the local community to rally round. A bell on the south turret served as a fog warning, and rooms were made ready for survivors who made it ashore. There was even provision made for coffins and funeral expenses.

This stretch of the northeast coast of Britain was also famous for its role in the development of the modern lifeboat. A local craft was adapted to withstand huge seas by a local man named Henry Greathead (1757–1818). He named his boat the *Original*, and launched her in 1790. This was not the first craft ever converted specifically to be used for saving lives. In 1785, Lionel Lukin (1742–1834), a London coachbuilder, had installed three large air tanks into a Norwegian yawl and attached a thick cork gunwale around the hull. A false keel of cast iron finished the impressive design off. Even earlier than Lukin, a Frenchman named de Bernières had been experimenting with lifeboats, but there is no evidence that any of his good ideas became a reality. Nevertheless, of all these so-called 'unimmergable' boats, Greathead's was the most favoured. He was granted £1,200 by parliament in 1802 and by 1824, 39 had been built. The world's oldest surviving lifeboat is one of those built by Greathead, the *Zetland*, which can be seen in the Lifeboat Museum in Redcar, just to the south of Newcastle. Also in 1824, William Hillary, a lifeboatman from the Isle of Man, who was appalled at the haphazard arrangements for lifesaving, founded the Royal National Lifeboat Institution (RNLI) – a national institution which drew together and co-ordinated the efforts of volunteers all over Britain. Not only did they declare their objective to be the assistance of all ships in distress, but also rewarded those who carried out rescues, and provided for the widows and families of those who died attempting to save others.

The history of lifesaving in the United States follows very closely the British pattern. Its origins, again, lay in the coastguard service. Established in 1790, it was responsible for both revenue collection and also the safety of lives at sea for which the revenue cutters were frequently employed. And as the 19th century dawned, the US coastguard service was faced with an unprecedented challenge in securing the safety of those who approached her shores, as the slow trickle of immigration with which she had always been faced, rapidly turned into an unstoppable flood, and hundreds of packed migrant ships approached her shores each year.

Lighting the Darkness

In these years of unprecedented attention to the wellbeing of seafarers, great advances were made in lighthouse designs. Lighthouse illumination had moved from wood to coal, pitch, oakum and then to candles until, in 1782, the Swiss scientist Aimé Argand (1750–1803) invented an oil lamp using a circular wick, through which was fed a current of air. Before long, his lamps poured out light from ten wicks fitted together, and were reflected by another new invention – the parabolic mirror; 600 glass mirrors set upon a plaster mould in the shape of a parabolic curve. Then, in 1822, the first dioptric lighthouse beam was introduced, which used lenses to collect and 'bend' the rays from a single source of light to project a parallel beam. To reduce light loss, these lenses formed a sort of cage around the light, and it was quickly realized that if this 'cage' could be made to revolve, then the light would flash. To help revolve this extremely large and heavy equipment, it was suspended in a bath of mercury. This was a highly significant breakthrough. In areas where there is one or more lighthouse, or other lights are visible from the shore, it is essential to be able to identify which lighthouse one is looking at by night. The original solution was to group different numbers of lanterns together, but this was impractical as it could easily cause confusion. The answer lay in flashing lights, each set to a different pattern.

The number and location of lighthouses also made significant breakthroughs in these years. In particular, it was the construction of lighthouses on isolated rocks and reefs that became the focus of maritime engineers. It was a considerable challenge: the building work could only be undertaken in the few hours that these rocks were exposed by the low tide, and usually no more than five hours' work per day was possible; the workmen then had to be transported to and from the rock safely, along with all of their heavy building materials. The sharp and always uneven surface of the rock or reef somehow had to be made level to provide secure foundations for the tower itself. Only then could specially designed interlocking bricks be fitted together in exactly the right order and orientation, like a giant jigsaw. The complexity of the construction was required for the lights to withstand the extraordinary destructive power of the sea in these isolated locations.

Three early attempts to build lighthouses on the submerged Eddystone Rock off Plymouth in 1698, 1699 and 1709 all ultimately failed, but by 1759 John Smeaton's robust tower stood there as an enduring beacon to mariners and an example to lighthouse engineers. The challenge was then taken up by Robert Stevenson (1772–1850) who in 1807 began work on the isolated Bell Rock off the north-east coast of Scotland. In these years it was Scottish ingenuity and energy that drove lighthouse design forward, as trade with America boomed, and the Atlantic winds brought transatlantic shipping quickly and easily to the wonderful natural harbours of Scotland.

On the other side of the Atlantic, the Americans also progressed in the protection of their mariners. In 1789, one of the first acts of the very first Congress centralized existing lighthouse building under federal control in 1789, and by 1852, 331 lighthouses, 42 floating lights, 35 beacons and over 100 buoys lit the American coast. French lighthouse building had also come under central governmental control from roughly the same period, and their achievements matched those of the Scottish. The Ar'men light, just off the Isle de Seine, itself an island detached from the most westerly point of Brittany, was built on a submerged rock. Not only is it exposed to the full force of the Atlantic, but is buffeted twice daily by six-knot tides.

Many of these great seamarks stand today. Built for mariners, the most remote are only ever seen by mariners, and yet they stand testament to the increasing awareness of land-locked politicians of the need to fund and coordinate the safety of their coasts. It has been estimated that in every year between 1793 and 1815 alone 2000 ships were wrecked, and in each of those years around 5000 British citizens perished as a result of shipwreck – a figure that exceeds the annual death toll on British roads in recent years. These statistics are in their own way quite shocking, but when they are set against the sheer quantity of voyages of any description that were made in those same years, then it is the number of ships that made successful voyages that is the most telling.

The *Queen Anne's Revenge*

May 1718

Long shrouded in myth – much of it self-generated – the pirate Edward Teach ('Blackbeard') captured the imagination of generations of people enthralled by tales of buccaneers. But in 1996, this stuff of legend was suddenly brought into focus by the discovery of what may well be the remains of Teach's flagship. Pirates' activities were clandestine and are largely lost to history, so the discovery of the Queen Anne's Revenge *would be a major find, helping us to piece together the extraordinary life of one of early colonial America's most colourful and notorious characters.*

Pirates survive in our collective consciousness because of their reputations as much as they do for the crimes that they actually committed. A successful pirate was a famous pirate and a famous pirate was a distinctive pirate. A pirate could be distinctive through cruelty, leniency, generosity, appearance, and gender, but in every instance a pirate's reputation had to precede him (or her). One method by which they increased the impact of their reputations was to cast themselves as evil, and they did so with a certain degree of relish and no little black humour. They came from hell and they were proud of it. Their flags signalled their intent with a variety of bones, skulls or sandglasses, a clear signal to their victims that their time was running out. All of this, of course, was designed to encourage their victims to give up without a fight; a sensible pirate would never risk his ship, his crew or his prize unless he had to.

The Devil Incarnate

To help achieve the necessary impact, one of the most well known of the pirates, Edward Teach (*c.*1680–1718), consciously cultivated his own image as the rebellious angel Lucifer himself. He plaited and braided his long black hair and beard and tied into it small fireworks that produced a glowing, fizzing, steaming stench: before he attacked your body, he attacked your senses. If you were lucky enough not to die at the hands of his crew, he made sure that you would never forget their captain, and the noise and smell that his crackling face produced. Unsurprisingly, his reputation grew even faster and greater than his success, and he was known by his nickname: Blackbeard. When William Bell stumbled into Teach's path off the Carolina coast he hailed the captain of the strange ship, asking him who he was and whence he came. Teach replied that he came from hell and he would carry him there presently.

Teach was one of a number of pirates who had profited from the War of the Spanish Succession (1702–13; also known as Queen Anne's War). In 1700 Charles II of Spain (r. 1665–1701) died and bequeathed all of his possessions to Phillip, duc d'Anjou, grandson of Louis XIV, king of France. The unification of both France and Spain under a single monarch was a terrifying prospect for everyone on the continent, but particularly Britain, whose interests lay in keeping her continental rivals divided. The British supported the claim of the Holy Roman Emperor to the Spanish throne and declared war on Spain. Much of that war was fought in Caribbean waters as the lucrative trade in silver, sugar and spices that originated in Spanish America and the Caribbean was the life blood of the war for both nations. During the war both sides sanctioned privateering – in essence, state-sanctioned piracy – against enemy trade. When the war ended many of those privateers stayed in the Caribbean, for it had everything that they desired.

There was an endless stream of valuable merchantmen forced to negotiate well-known passages through the necklace of reefs and islands that decorate the Caribbean. Those reefs and islands, moreover, provided an almost limitless network of hideouts and lairs from which they could work. The lively illicit economy, flooded as it was with merchants, seamen, slaves and a great deal of money, provided the perfect opportunity to sell quickly any captured goods. And finally, but perhaps most importantly, any significant naval presence was temporary; in the early 18th century not one nation with a substantial navy had a sufficiently established naval infrastructure to support a squadron in the Caribbean for any lengthy period of time.

Teach was one of those men who saw the opportunity and was prepared to take the risk, and he did so with great success. They took piracy to a level that would interrupt international trading patterns and would eventually force a violent reaction from the European navies. The first stage in that reaction came in 1718 when Governor Woodes Rogers (*c.*1679–1732) established royal authority in the Bahamas, which had become the main pirate rendezvous. The pirates scattered; many to the west coast of Africa around Sierra Leone, where there were easy pickings to be made and little threat from authority, and many also to the coast of Virginia, which had grown fat from the Caribbean trade. The pirates who chose that path could happily lurk in the unpopulated inlets of the Carolinas. By 1718 Teach had made that coast his hunting ground, and his alone, patrolled by his flagship the *Queen Anne's Revenge* and three sloops. Perhaps tired of plundering ships, Teach decided to plunder the source of their wealth, and so in May he blockaded the town of

Charleston itself, looting every ship in the harbour and holding the town to ransom.

Blackbeard's Violent End

Teach's fearsome reputation and audacity created his success but it was also the source of his downfall and on the morning of 22 November, 1718, he and his crew were surprised by a naval force that had been directed to his location, just off North Carolina, by some local merchants who were exasperated by his blatant thieving. They had complained to Governor Alexander Spotswood of Virginia (c.1676–1740), a man with a hatred of pirates that coursed through his very veins and who cherished the prospect of ridding the seas of their kind. Also a canny politician, he knew that a big scalp would have a greater effect than the removal of a single pirate, and the capture and execution of Blackbeard would reverberate through the waters of America.

Teach was found drunk and asleep with his crew aboard the sloop *Adventure* in the lee of Ocracoke Island, part of the outer banks of North Carolina, southwest of Cape Hatteras. They fought like men who knew that their only chance was escape; there was by now no mercy shown to a convicted pirate, and Teach's crew nearly overcame the combined naval force of two sloops. The British force was led by Lieutenant Robert Maynard, who had carefully hidden many of his men below decks. Just when Teach thought he had won, the sailors swarmed up on deck and took the pirates by surprise. Teach died in the fight with five shots in him and over 20 sword cuts. The death blow came from a Highlander in Maynard's crew, who severed Blackbeard's head.

Teach's extraordinary death is fitting for the extraordinary life that he led, and his fearsome reputation has survived in generations of books and films about pirates. More recently, however, historians have started to research the real Teach in an attempt to strip away the fiction from the fact. There are numerous references to his activities in a contemporary newspaper, the *Boston News-letter*, and there are further references from the trials of a number of pirates in 1719. A number of letters from British naval captains and other British officials in the Caribbean and East Coast of America who came across Teach, survive today in the British National Archives, and there is similar evidence from French authorities in Martinique, who heard of Blackbeard's deeds. Charles Johnson's timeless *A General History of the Robberies and Murders of the Most Notorious Pirates*, published in 1724, only six years after Teach's death, provides a great deal of background information to the era and some specifics relating to Blackbeard. But these sources are few and, as with many histories of such characters who live beyond the reaches of ordered society, whose life and activities are illicit, much of

Teach's life remains intangible. In spite of this research he is still an enigmatic character of dappled light and shade reflected by the patches of our knowledge and ignorance. In 1996, however, our knowledge of Teach and his crew took a giant leap forward, a leap that cements this tale – which at times is scarcely believable – into the pages of history: a team of archaeologists actually found his ship, the *Queen Anne's Revenge*.

Pirate Ship Extraordinaire

Undoubtedly, part of Blackbeard's reputation stemmed from his ship. Most pirates captained small, fast sloops, schooners and galleys, but only a handful had vessels of a significant size. The *Queen Anne's Revenge* was said to be large enough to carry 40 guns and was manned by a crew approaching 300 men, whereas a ship capable of carrying 10 or perhaps 20 guns and a crew of 150 or 200 seems to have been the average for a pirate. A 40-gunner would easily outmatch the Royal Navy's smallest ships of the line, and the only pirates who could claim to match that strength were Bartholomew Roberts (1682–1722), William Moody and Henry Avery (b. 1653).

Teach had captured her in the Caribbean in the winter of 1717. We know from the report of Henry Bostock, captain of a sloop Blackbeard captured a month or so later, and who was kept on board his ship for a full eight hours before being released, that she was probably Dutch-built and was a French Guinea Man. Historians have since discovered that she was indeed French, and that she was originally the slaveship *Concorde* owned by René Montaudin of Nantes at the mouth of the Loire, the centre of the French slave trade and one of that town's most successful slave traders. A typical journey saw her sail to Juda, on the west coast of Africa, where she would load her human cargo, usually between 300 and 500. She would then sail to the Caribbean to trade at Guadeloupe, St Domingue or Martinique before returning to France. Her voyage that began on 24 March, 1717 was typical until Teach seized her off the island of St Vincent. The slaves aboard were taken to Martinique and sold before Teach resumed his pirating aboard his new ship, which he named *Queen Anne's Revenge* in open defiance of the world ruled by that British sovereign, which he and his men had left behind.

Success followed success for Teach's formidable new ship, and she was soon the flagship of a small fleet that included three sloops and, by May 1718, as many as 700 men. This was the fleet of ships that blockaded Charleston in the spring of 1718. Soon after their successful raid, Teach took his crew north to Beaufort, North Carolina. There, and perhaps intentionally to break up the pirate crew after the raid on Charleston that would surely bring the full force of the navy to the hunt, the *Queen Anne's Revenge* ran aground and foundered on the outer bar.

There she lay until 1996 when archaeologists located, mapped and began to excavate the wreck. Within four years of summer excavations, over 2000 artefacts had been recovered. Ship parts and equipment, arms, scientific, navigational and medical instruments, food preparation utensils, storage containers and many personal items have all been raised, even a small quantity of gold dust. The remains suggest that she sank with her bow towards the shore, when she heeled onto her port side and spilled her contents onto the seabed. A great anchor from the same period lies around 130 metres (420 ft) to the south and has been interpreted as a final desperate attempt to haul the ship clear of the sandbank.

There is, however, some debate over the identity of the wreck, and there are those who believe that the association of the wreck with Blackbeard is perhaps a little premature; there is, after all, no indisputable evidence that the wreck is indeed the *Queen Anne's Revenge*. At the same time, though, there is no evidence at all – and this is from a site that is so well preserved that her excavators estimate that she might yield as many as a million artefacts in the coming years – that she is *not* the *Queen Anne's Revenge*. The artefacts so far recovered certainly all point to the right date, and the loss of such a large ship from that period would, without any doubt, have produced some trace in the official records. But there is none, and this suggests, perhaps more strongly than anything else, that it is the remains of a ship that plied its trade outside the law. The remains, moreover, strongly suggest the story of a ship that simply ran aground on the sand bank and then sank *in situ* – exactly what we know to have happened to the *Queen Anne's Revenge,* in exactly that location, at roughly that time. It is also documented that when Teach first captured her, she was carrying a cargo of 7 kilograms (20 lb) of African gold dust. The discovery of gold dust at the wreck site has therefore become highly significant, and scientific analysis has proved that gold of the size and shape discovered at the wreck site does not feature naturally around the North Carolina coast.

If the identity of the *Queen Anne's Revenge* is accepted – and the consensus of historians and archaeologists is certainly leaning that way – then it is a significant find indeed, as only one other vessel has been positively identified as a pirate ship: Sam Bellamy's *Whydah*, discovered off Cape Cod in 1984. In the coming years, therefore, these finds will open a window into life aboard ship in this golden age of piracy; a peculiar life beyond the fringes of accepted society that was governed by passionately held beliefs of code and honour, and bound together by a common respect and a sense of fraternity among all of those who had rejected the protection of their own nation's sovereign. Now they fought together under the common banner of King Death, known typically by those pirates who found humour in the blackest of acts, as the 'Jolly Roger'.

The Battle of Quiberon Bay

20 November 1759

Edward Hawke's attack on a squadron of French ships in Quiberon Bay during the Seven Years' War was one of the boldest strikes in naval history. Undaunted by treacherous shoals and reefs, Hawke made superior seamanship tell against the enemy fleet. At the end of the engagement, seven out of 21 French ships of the line had been destroyed – with heavy loss of life – while the British commander lost just two of his blockade force of 23, their crews rescued.

None of the wrecks featured so far were the direct result of enemy action. It is true that there may have been spear-heads embedded in the hull of the Kyrenia ship (see pages 43–5), and that *La Trinidad Valencera* (see pages 90–5) was forced to seek shelter in Northern Ireland because of damage sustained during Drake's fight with the Armada, but in both cases that damage did not lead directly to the sinking of either ship. This might seem surprising, but it is in fact indicative of a very important characteristic of wooden sailing ships: they were extremely difficult to sink. In addition, they also represented such vast quantities of investment in time and money that it was rare indeed for an attacking ship to desire to sink her opponent. The ultimate aim was always to secure her as a prize. Not only did a captured ship make the captain and crew of the victorious ship a fortune, it was also a powerful symbol of military success to be flaunted in the face of the enemy. Moreover, the capture of an enemy ship had twice the impact of that ship's destruction: not only did the enemy nation lose a ship, but the victorious nation gained one.

Tactics in the Age of Sail

The type of damage inflicted on one ship from another depended very much on the circumstances. A pirate or a privateer would try to damage her target as little as possible to maximize the potential reward, and in these circumstances they would target the rigging of their opponent to disable her. Similarly, in chase, both ships would target the rigging of their enemy. Once an engagement had been forced between single ships or fleets, the type of damage inflicted depended almost entirely on the range at which the battle was fought. When some distance away, perhaps at 1000 metres, it was still possible to target the enemy rig, but when bilge to bilge no such subtlety of aim was possible, and both ships simply poured their fire directly into the enemy hull, perhaps alternating the aim of each tier of guns up or down so as to catch the enemy crew in a crossfire. But even in these situations, the enemy gun crew was the real target, and not the ship itself; so much so that the charge of the cannon was often reduced when engaging at close quarters so as to reduce the speed of the cannon ball. A ball travelling at high velocity would simply punch a circular hole through the enemy hull, which could quickly and easily be repaired by pre-prepared plugs lined with felt, tallow and tar. A shot travelling more slowly, on the other hand, had a devastating effect on a wooden hull, as the shock wave that travelled through the wood shattered the walls of the gun-deck and sent foot-long splinters among the guncrews in a whirlwind of death. The damage reports of warships involved in fleet battles are therefore full of descriptions of ships with hundreds of shot lodged in their hulls – rather than hulls perforated like a colander.

Wooden ships, moreover, were essentially organic: they were made entirely out of wood, canvas, rope and to a lesser extent, iron. All instances of damage could be repaired given time. Ships carried a great deal of spare timber that could be used to fashion almost anything, from a spare rudder to a new bowsprit, as well as specific spare parts such as masts and yards. In fleets there was also a generous exchange of parts and labour to help those in distress, and in most instances even a terribly badly damaged ship could be patched up sufficiently to limp to a friendly port. It is characteristic of the era, therefore, that very few ships were actually wrecked in battle. There are of course some exceptions, where heavily damaged ships could not be kept free of water, or where fire rampaged through the gundecks and found its way to the ship's magazine, but these were isolated cases. Against this background the Battle of Quiberon Bay in 1759 stands out, for on that dreadful night eight huge warships were wrecked and nearly 3000 men died.

Sworn Enemies

Britain and France had been at war over the fate of Austria from 1740 to 1748, when a tentative peace was reached, but by 1755 they were at war again, having failed to agree the demarcation between British and French possessions in North America. The ensuing conflict, the Seven Years' War, was the first to be fought around the globe. Prussia and Hanover allied with Great Britain while Austria, Russia, Saxony, Sweden, Spain and Portugal fought with France. However, only Britain and France contested control of the world's oceans, and while naval strategy centred on attacking each other's colonial outposts, the French also harboured invasion plans for England itself, and began to assemble a huge invasion force around the southern coast of Brittany. To make the crossing, the French army needed a vast fleet of transport ships, and for a limited time, control of the Channel. Neither of these could be achieved with the French navy in port: they first needed to collect the transport ships from a nearby bay known as the Morbihan, within the great bay of Quiberon, and then run the gauntlet of the Royal Navy.

The British, meanwhile, were keenly aware of these plans, and so had instigated a close blockade of Brest, the main French naval base on the Atlantic coast. The blockading force was part of the Royal Navy's Western Squadron, a standing force that patrolled the Western Approaches to the English Channel, which was under the command of Edward Hawke (1705–81), one of the most gifted and aggressive admirals of his generation. Moreover, the ceaseless blockade in the challenging waters of Biscay had honed the British seamen to an exceptional state of professionalism while the French had languished in port. British sailors, at least, were sure of their superiority over the French, but at home the war was taking its toll on the national psyche. It was well known that France's grip on its colonies was weakening and that it planned a final throw of the dice to change the course of the war; the gathering invasion army cast a long shadow over England.

Daring Manoeuvres

Hawke's chance to thwart the invasion came in November 1759. The winter gales of unceasing westerly winds threatened to blow the blockading force against the treacherous coastline of Brittany, which was guarded by semi-submerged rocks and reefs and shrouded in fog. To keep safe in such a sea in a single ship was a challenge in itself, but to do so with a fleet was almost impossible due to the heightened risk of collision. In such conditions the Western Squadron sought shelter in Torbay. As the weather eased on 14 November, Hawke set sail to resume his blockade. He soon learned that the commander of the French fleet at Brest, the comte de Conflans (1690–1777), had taken advantage of the curious weather – a storm followed by a calm, with light easterly winds – to break out of port. The chase was on.

Hawke knew they would head for Morbihan and he set all sail to catch the French fleet, which was almost 200 miles (320 km) away. To catch them, Hawke needed a stroke of luck, and it came that day in the form of a howling easterly gale that blew both fleets deep into the Atlantic. This gave Hawke the extra time he needed for the superior British seamanship to tell. By the early hours of 20 November, both fleets were off Belle-Isle, the island that sheltered Quiberon Bay. Just after 7 o'clock a scouting frigate positioned on the horizon used her sails to signal that she could see an enemy fleet. The British quickly and efficiently formed a line abreast while the seven closest ships set all sail and chased. Hawke knew he had limited time and even less sea-room. The days are short in the Atlantic winter, and in November the Biscay sun sets before 4 p.m. The weather had also whipped itself into a fury and was blowing hard directly onshore. Conflans took solace in both of these factors and headed for Quiberon Bay.

The bay itself is guarded by large islands, rocks and shoals, and its entrance is only 6 miles (10 km) wide. Once inside the outer bay, another channel, only 2 miles (3.2 km) wide, protects the inner anchorage. To thread a large fleet of wooden warships through such a small gap, in fading light and in a storm, was difficult enough with experienced local pilots on board, but to attempt to do so with no knowledge of the local hydrography would be suicide. Conflans reasoned, accurately enough, that the British captains did not know the area well, and that he would be safe in the sheltered waters of Quiberon Bay. But he did not reckon on Hawke's tenacity. Hawke knew that the stakes were high and that he dare not overlook this opportunity to strike a decisive blow against the French fleet and their planned invasion. He was also confident in British seamanship, and he solved the problem of unknown hydrography as any good sailor did in a chase in the Age of Sail: he reasoned that wherever his enemy could go, so could he. While the French ponderously negotiated the entrance to the anchorage, Hawke kept all sail set and hurled his ships at the French rear.

By all accounts, Conflans was dumbstruck. The one thing he and his crews did not expect was for Hawke to follow him with his entire fleet, but now, as well as avoiding the surrounding rocks in the stormy weather, they also had to fight the British ships, ferocious as wild dogs. The British had spent endless gruelling and boring months simply blockading Brest with rarely a glimpse of their enemy, but now they had seen them, they had chased their quarry down, and they were consumed by bloodlust.

An Exceptional Engagement

The action that followed was one of the most chaotic of the whole Age of Sail. Ships of both nations fought the weather as much as they fought each other. To engage in such conditions was always very dangerous. By 1759, warships were much more stable than they had been previously, but a sudden squall could easily force a warship to heel quickly and submerge her lowest gunports. While the guns were being fired, therefore, a close eye had to be kept on the weather, and the sheets of the sails all let fly to release the pressure if a squall hit. It was one further dimension to the already formidable challenge of sailing warfare, and the French, who had not been at sea for months, were simply not up to it. In the midst of the battle a strong gust of wind blew the French 74-gunner *Thésée* flat, and water poured into her gunports. She sank with appalling loss of life, with only 22 of her crew of more than 600 surviving.

To lose an 18th-century warship in this way and in the heat of action was rare indeed but the Battle of Quiberon Bay was exceptional in many ways, and shortly after the *Thésée* went down there was another remarkable incident. Hawke came alongside the French 70-gunner *Superbe* and poured two broadsides straight into her hull. This was usually only enough to maim large numbers of the crew. The hull, although damaged, would still retain most of its structural integrity. But in this case, and it is likely that this is a unique occurrence in the Age of Sail, the *Superbe* immediately sank with all 800 of her crew. More disasters followed. Two other French ships (including Conflans' flagship *Soleil Royal*) were driven ashore and deliberately burned to avoid capture, and two more were wrecked on the reefs and sandbars of the bay. Two British ships were also wrecked, both on the Four shoal, which lay 5 miles (8 km) to seaward of the bay, and was unknown to British pilots.

The Battle of Quiberon Bay has no parallel in naval history. In no other battle were so many ships wrecked, and never was such a bold decision taken as that by

Hawke when he resolved to chase the French against a lee shore in a rising gale as night fell. In this case the success justified the risk. The strength and determination of the French navy was shattered and the invasion threat lifted. British naval superiority remained unchallenged until the War of American Independence in 1775. One contemporary summed it up by declaring '*The glory of the British flag has been nobly supported, while that of the enemy is vanished into empty air*'.

La Méduse

17 July 1816

One of the most infamous names in maritime history is that of Hugues Duroy de Chaumareys, captain of the French frigate La Méduse, *who abandoned survivors of a shipwreck caused by his own incompetence to a nightmarish ordeal of bloodshed, starvation and cannibalism. When the gruesome story broke in France, it sent shock waves through a nation already crippled by self-doubt and deep social divisions after the defeat of Napoleon. The plight of those on board the raft of the* Medusa *was immortalized by the painter Théodore Géricault in a monumental and highly symbolic canvas that can be seen hanging in the Louvre in Paris.*

On 17 July, 1816 off the coast of Senegal, officers aboard the French brig *L'Argus* saw something floating in the water. They hoisted all sail and rapidly closed, and presently made out a raft with a makeshift sail of tattered canvas, and littered with bedraggled figures. An officer of the ship, M. Lemaigre, launched a boat and rowed across. According to one account, what he found on the raft were men *'lying on the boards, hands and mouth still dripping with the blood of their unhappy victims, shreds of flesh hanging from the raft's mast, their pockets filled with these pieces of flesh upon which they had gorged themselves'.* The *Argus* had chanced upon the raft of the *Medusa.*

The story behind the raft is one of the most complex in the whole history of shipwrecks, for it became far more than just the loss of a ship. The story was used by politicians, activists, playwrights and artists as source material for political and social commentary, and it has endured as one of the most famous shipwrecks of all time due to its immortalization on canvas by the young and exceptionally talented French Romantic painter Théodore Géricault (1791–1824), an artist determined to make his name through a grand and provocative painting. Indeed, the wreck became famous because – for a whole variety of reasons – the story behind her sinking was itself so provocative.

France's New Order

Just a year before the *Medusa* was wrecked, Napoleon had been defeated at the Battle of Waterloo and exiled to the tiny island of St Helena, deep in the South Atlantic. Napoleon's once great empire, which had dominated so much of the continent, now lay in ruins. British maritime strength was unchallenged and unchallengeable. France was surrounded by powerful nations she had made her enemies over the preceding decade. There was, moreover, little appetite for bold leadership; Napoleon had time and again demonstrated the pitfalls of power residing in the hands of one man. To make matters worse, the Bourbon monarchy that had replaced Napoleonic rule was anxious both to reward its followers and to overlook those who had fought under Napoleon. French society was still bitterly divided and incapable of acting in unison for the good of the country.

One of the areas where this was most visible was in the appointment of men to military positions. Once the loyalty of the army and navy could be assured, the Bourbons reasoned, then the possibility of yet another uprising in favour of the ousted emperor was far less likely to occur. The result was the appointment of men to positions for which they were either unqualified or hopelessly out of practice. One of those men was Hugues Duroy de Chaumareys (b.1763), an aristocrat who had not served in a French ship for 20 years, but who nevertheless was appointed captain of the frigate *La Méduse*.

Senegal was first settled by Portuguese explorers who circumnavigated Africa in the late 15th century. It was not long before they started to trade slaves, and for more than a century the region around the Senegal and Gambia rivers was the largest supplier of slaves to Europe. It was soon overtaken in importance to the slave traders by more densely populated areas of central and western Africa, but the region of Senegambia retained a significant reputation, and still exported thousands of slaves each year. As a site of commercial interest it inevitably attracted warfare and changed hands a number of times. The French, who had arrived in 1628, were driven out by the British in 1758, but it was restored to France in 1783. Again, during the Napoleonic Wars it was taken by the British, and once more restored to France after the war. But that final exchange, in which Senegal was returned to French control, was loaded with far more significance than it had been in 1783. Britain, now the undisputed dominant world power, had decided in 1807 to make a stand against slavery and so had abolished the practice in all of her colonies. For them, therefore, the value of Senegal had greatly been reduced. The British, moreover, were not content with setting a lone example but were determined to carry everyone with them on their crusade to eradicate the trade in African slaves.

The *Medusa* fits into this story because she formed part of a small convoy of four French ships (also comprising the *Écho*, *Argus* and *Loire*) that was dispatched on 17 June to the port of Saint-Louis in 1816 to restore French control of Senegal; the *Medusa*'s passengers included a number of dignitaries, among whom was the new governor, Julien

Schmaltz. Once he had landed, control of that entire area was to be handed over to him by the incumbent British governor, and that transition of power carried with it one crucial proviso: that the French did not use the colony to trade in slaves. Once in power, however, Schmaltz would clandestinely continue to trade slaves while maintaining a public façade that supported abolition. This, therefore, was the context within which *La Méduse* had sailed, already so politically charged that it could become news in itself, but the events that followed were to secure her place in history.

Abandoned to their Fate

Somewhere in mid-Atlantic, Chaumareys committed the cardinal maritime sin of losing touch with his convoy, and as he neared the shallow waters of the West African coast, riddled with sandbanks, this was nothing short of a disaster. Tragedy could have been averted if he had kept his nerve, but it failed and he meandered *La Méduse* erratically through the sandbanks until she ran aground on the Arguin Bank on 2 July. Numerous efforts were made to haul her off but she was stuck fast. She had also struck at the very top of one of the highest tides of the year: there was no chance that nature would refloat her, and she had to be abandoned.

It was clear that the 400 souls aboard the *Medusa* would

not fit into her six lifeboats, but the situation was calmed by Schmaltz's proposal to build a raft to carry the provisions and the extra men, which would be towed behind the boats. Hastily constructed of spare masts and yards and pieces of deck, it was almost 28 metres (65 ft) long and 8.5 metres (28 ft) wide. The provisions, including huge casks of alcohol, were stowed, and a portion of the *Medusa*'s passengers allocated to the raft. But the division was clearly unfair and certainly not sensible. The captain and all of the senior officers took to the boats, leaving 150 men on the raft, with no authority and precious little maritime experience. The raft, moreover, submerged under their combined weight, one survivor recalling '*it had sunk at least three feet, and so closely were we huddled up together that it was impossible to move a single step. Fore and aft, we had the water up to our middle.*' This sorry fleet set off for the nearest point on the wild African coast, some 50 miles (80 km) away. For many, this held as much terror as their present predicament; the Senegal coast was renowned in popular fantasy for the savagery of its inhabitants, especially the Muslim Moors, whom French Christians thought lusted after human flesh.

It is unclear exactly what happened next and why, but by daybreak the raft had been abandoned by the boats. They had drifted within only 12 leagues of the shore and it is likely that upon discovering themselves so close to safety, the boats cut the lumbering raft free and sped to shore. Soon after, an offshore breeze sprang up and the raft disappeared over the horizon with its cargo of leaderless, incompetent and terrified humanity, surrounded by alcohol and armed to the teeth.

Overcome by despair they drank, and then they fought, with guns, knives, axes and swords. Many died in this open battle and many others survived with hideous open wounds. They began to drink their own urine and seawater, and some went mad. Others began to eat the dead. By the sixth day, of the original 150, only 28 survived. Of those 28, the stronger realized that there were insufficient rations for everyone, and so the sick and wounded were tossed overboard. After this purge only 15 remained, who survived for another week before they were saved by the *Argus*.

When the story of the wreck first reached France, the government immediately appreciated that it was a potential political bombshell, and did their best to suppress it. The following terse announcement appeared in a national newspaper:

> *'On 2 July, at 3 o'clock in the afternoon, the frigate* Medusa *was lost, in good weather, on the shoals of the Arguin 20 leagues distant from Cap Blanc. The* Medusa's *six launches and lifeboats were able to save a large part of the crew and passengers, but of 150 men who attempted to save themselves on a raft, 135 have perished.'*

The Scandal Erupts

There the matter rested until two of the raft survivors – Henri Savigny, the *Medusa*'s surgeon, and Alexandre Corréard, a geographer and engineer – made their way back to France. Savigny's narrative was sent to the Minister of the French Navy, but it was deliberately buried. Another copy soon found its way to the hands of Élice Decazes, Prefect of the Police and a man of unquenchable political ambition. He knew the value of the account and immediately leaked it to the anti-Bourbon newspaper, the *Journal des Débats*. Four days later it was published in *The Times*. The uproar that followed brought so many simmering resentments to a head that all of Europe became gripped with the fate of those men so callously abandoned off the African coast.

Political opponents of the monarchy used the story as a broad attack on the government; they represented the loss of the *Medusa* as a political crime rather than a natural disaster, and blamed the minister who had appointed the incompetent captain. The wreck became an illustration of the danger to which France was exposed by a regime that put dynastic over national interest; that gave command of ships to political favourites, and allowed aristocratic officers to abandon their men in times of crisis. Through the prism of such liberal politics, the story of the raft became that of France herself, adrift, threatened on all sides, and consumed by violent internal struggle.

Interwoven with these political issues were those of slavery, and they were also grasped firmly by critics of French policy. Alexandre Corréard, one of the two men responsible for the most detailed first-hand accounts, was a passionate abolitionist and used his account to condemn the continuation of French slavery. At the time that Corréard's enlarged second edition of his narrative was published, now swollen with invective against slavery, the young artist Théodore Géricault picked up the story.

In many of the careful sketches that Géricault executed in preparation for the final work, and also in its final version, a strikingly handsome black man is at the very apex of the composition, raised above his fellow survivors, frantically signalling for help to a distant ship on the horizon. He is generally believed to have been Jean Charles, a black soldier and survivor of the raft, and his inclusion by Géricault is highly significant. He is a symbol of hope; to save him is the only means by which the raft – and France – can be saved.

There is, however, one final twist to this issue of race on the raft, and it has to do with cannibalism. Without any doubt, the account of cannibalism shocked Europe more than any other part of the story. Civilized Europeans did not eat each other; indeed the accounts of Corréard and Savigny were very careful to point out that the few officers aboard the raft were the last to eat human flesh; they certainly did not instigate it, and when the officers did become involved, Corréard and Savigny suggested that the flesh was first dried in strips to make it less disgusting. Other accounts claimed exactly the opposite; that the officers were the first to begin eating the men. This provided even more ammunition for those opponents of an aristocratic and class structure that could allow such events to occur, and it rang bells of other shipwreck stories in which social strata had a direct bearing on who was eaten first. Among British shipwrecks in the 18th century, for example, the first victims to be eaten were often the black slaves, swiftly followed by the Portuguese and the Spanish. The black man atop the raft, therefore, is not just saying '*save me*' – he is also saying '*save me* ***from them***'.

With the French so embroiled in this public self-destruction, it was perfect fuel to the British flame, ever-present in those years, of national superiority. To many British, the disaster aboard the raft clearly occurred because they were French; that in itself was enough to explain the chaos and immorality of their behaviour. Indeed, these sentiments were further fuelled by the wreck of a British naval ship in exactly the same year, which demonstrated the British character in adversity in the finest possible light. When HMS *Alceste* ran aground in Indonesia, a similar raft was made, but this time only for supplies. Order was maintained in the boats as the weather, the sea, thirst, hunger and Malay pirates were all fought off, with all survivors following the personal example of Lord Amherst, British Ambassador to China.

Never before the *Medusa* wreck, and never since, has a shipwreck been embroiled in such a tangle of issues. What started as a simple navigational error became a *cause célèbre*, and as the *Medusa* found herself stranded, isolated and alone, so too did France as the story broke. Her fall from those mighty days when Napoleon led a magnificent empire was complete.

The *Essex*

20 November 1820

The factual basis for Herman Melville's masterpiece Moby Dick *(1851) was supplied by the almost incredible tale of the Nantucket whaler* Essex. *While out hunting in the South Pacific in 1820, midway between the Galapagos and Marquesas Islands she was deliberately rammed by a sperm whale and rapidly sank, leaving only her whaleboats to transport the crew to safety. Shunning landfall on the nearest islands for fear of being eaten by cannibals, they were eventually forced to turn cannibal themselves as they sailed across the vast expanse of ocean that separated them from the South American mainland.*

It is a peculiarity of male sperm whales *(Physeter macrocephalus)* that the posterior six cervical vertebrae of their spines are fused together. This mass of fused vertebrae facilitates the transfer of energy from an impact at the head to the body, reducing the risk of spinal compression injury. Moreover, the foremost part of its skull is reinforced with transverse partitions and thick fibrous tissue, and is perfectly aligned with the cervical vertebrae. The skin that covers the skull is also unusually thick and tough and the tail can propel the great mass of whale, weighing up to 80 tons, at speeds approaching 6 knots. In short, the male sperm whale is a perfectly designed battering ram. Although it has not been proved definitively, it is thought likely from their physiology that the head of the sperm whale has evolved as a weapon, to be used in male–male aggression, usually during competition for females. It is an activity that is also displayed in humpback whales, bottle-nosed whales, Amazon River dolphins, narwhals, long-finned pilot whales, bottlenose dolphins, spotted dolphins and killer whales. In that respect, therefore, the sperm whale is not unique, but among the crews of those ships who for centuries plied the oceans to hunt these giant creatures, the sperm whale was renowned for its aggression, and it was feared.

In those days, whales were harpooned from open boats, no more than 9 metres (30 ft) long and about 2 metres (6 ft) wide. They were launched from the whaling ship, which served as home for the crew and a sort of factory where the whale carcass would be processed and reduced to its commercially valuable parts of oil, blubber and ivory. Once harpooned, the whale would be forced to tow the boat – or boats – in an effort to escape. It would soon tire and could then be killed more easily, stabbed in the lungs by a giant lance until it drowned in its own blood. The fragility of those boats alongside a sperm whale is strikingly obvious, but the sailors knew that a whale might even be capable of sinking the whaleship itself.

The whaling industry was notoriously isolated, however (as it still remains), and sailors of all types were renowned for their tall tales of adventure and daring on the high seas. Taken together, therefore, the general public knew very little indeed about the whaling trade during these years, and few would believe a salty tale of a ship sunk by a whale, let alone an apocalyptic warning that such a disaster could happen at all. But all this changed in 1821, when two men were found off the coast of Chile in a whaleboat, surrounded by piles of human bones. Their tale was scarcely believable. They claimed they had been at sea for 94 days and had sailed over 4500 nautical miles (8325 km) – that is 500 miles (926 km) further than Captain William Bligh's more famous open-boat journey, made after he was cast adrift by mutineers from the *Bounty* in 1789. But the men in this whaleboat had not been cast adrift by other men, nor had they lost their ship through human error or by the force of wind or sea; their ship the *Essex* had been sunk, deliberately, by a whale.

Enshrined in Fiction

The story of the *Essex* instantly became famous throughout America, but was soon lost to posterity, being replaced by that of another ship, the *Pequod*, the invention of a young author named Herman Melville (1819–91). Melville used the fate of the *Essex* as the climax of *Moby Dick*, his fictional tale of life, death and revenge on the high seas. Melville himself had considerable first-hand experience of seafaring. Born of excellent stock, he fiercely resisted all attempts to make him lead a conventional life along traditional routes of employment made cosy by an established network of protection and patronage. He explained to a friend: *'For my part, I abominate all honourable respectable toils, trials and tribulations of every kind whatsoever.'* To prove it, he left the comfort of his family home to make his way as a penniless, common sailor. Melville served in merchantmen, whaleships and men of war; he experienced mutiny and lived with practising cannibals in the Pacific for four months. On his return, he resumed the comfortable life he had left before, and married the daughter of the chief justice of Massachusetts. Aged 25, and having already seen so much, Melville considered that his life could begin, and he settled into a literary existence that drew extensively on his knowledge of the sea.

There is no doubt that the climax of his achievements is *Moby Dick*. A sophisticated, multi-layered allegory of vast scope, it is littered with lengthy and detailed descriptions of the whaling industry which serve, in part, to legitimize the story. Whilst working on the book he wrote to a friend:

> *'It will be a strange sort of book, tho', I fear … to cook the thing up, one must need throw in a little fancy, which from the nature of the thing must be ungainly as the gambols of the whales themselves. Yet I mean to give the truth of the thing, spite of this.'*

He did so, and the passages on whaling in *Moby Dick* remain among the fullest, truest and most readable descriptions of that trade that have ever been penned. His research on the fate of the *Essex* was also meticulous. *'I have seen Owen Chase who was chief mate of the* Essex *at the time of the tragedy,'* Melville wrote, *'I have read his plain and faithful narrative: I have conversed with his son; and all within a few miles of the scene of the tragedy.'*

Nevertheless, the setting was fictional and the factual basis of the climax to his story was soon forgotten. Melville died in obscurity in 1891 and it was not until long after his death that his reputation blossomed as one of the 19th century's finest writers. As his reputation grew, however, so his story of the *Pequod* became more famous, and thus our knowledge of the aggressive capabilities of sperm whales became popular currency. As the focus of critics shifted to the formidable (and ongoing) challenge of unpicking the layers of meaning within *Moby Dick*, so was the background to his writing and his influences analysed, and soon the story of the *Essex*, once again, came to be retold.

The Fate of the *Essex*

Accounts by two survivors from the wreck of the *Essex* have provided us with a remarkably detailed and balanced account of what happened. The *Essex* had been hunting deep in the Pacific, where whalers had been forced to seek their prey by the success of generations of American and European hunters. After some weeks with no kills, they came across a shoal of whales and took to their boats. First mate Owen Chase was in command of one, and as they approached the area where the whales had been sighted, one surfaced, directly under the boat, and smashed it, casting its crew into the sea. Chase was forced to paddle the wreckage back to the *Essex* to repair her: they had already lost a number of boats, and it was essential that this one survived.

Meanwhile, two whales had successfully been harpooned and the sailors' fortunes seemed to be turning. But while Chase was repairing his boat, he looked up and saw a bull, perhaps 26 metres (85 feet) long, lying very still in the water 100 metres (320 ft) or so from the ship. He had the distinct impression that he was being watched. The bull suddenly dived, before surfacing only 30 metres (98 ft) from the side of the ship and then charging it. The whale struck amidships and the *Essex* shook as if she had hit a rock. The whale then swam away to leeward where it thrashed about, seemingly enraged, before launching another attack, this time at the ship's bow. On this attack, the whale's speed was twice that of before, and the wake made by its tail almost 12 metres (40 ft) across. It stove in the hull just under the cathead. Within ten minutes the *Essex* had capsized.

When the *Essex* sank she was already 20 years old, but she had been made out of beautiful American white oak, a species that grows to an enormous size. The sheer size of the limbs of these trees, which elsewhere would seem as trunks, allowed shipwrights to fashion crucial structural parts of a ship out of single pieces of timber, and the fewer joins that there were, the stronger a ship would be. The density of the wood itself also made it almost impervious to rot. The ship's ribs were made of timbers a foot square and over these were laid planks – again of white oak – 10 centimetres (4 in) thick. These in turn were covered by planks of yellow pine more than 1 centimetre (half an inch) thick. This was then sheathed in a thick layer of copper to protect her hull from the boring teredo. The

Essex may have been two decades old but she was still incredibly strong.

In complete disbelief – a state of '*vacant idleness*' – the men wandered the wreck of the *Essex* until they were sufficiently lucid to form a plan. Three whaleboats were left for 20 men, and they were carefully rigged, provisioned and their sides built up to withstand the ordeal that undoubtedly lay ahead. As the preparations were made to abandon the ship, they drifted with the current and south-easterly breeze further away from the South American coast – 50 nautical miles (92 km) in just one day. The nearest land was the Marquesas Islands, 1200 miles (2220 km) to the westward, but their reputation among the Nantucket whalers for cannibalism and ritualized homosexuality instantly ruled them out. Other islands to the westward had similar reputations and so it was decided to head due south. Around 1500 miles (2775 km) away was an area of the Pacific that was known for its westerly winds, and these, it was reasoned, would blow them the 1000 miles (1600 km) back to the South American shore. In dread of the Pacific natives, these men therefore committed themselves to a journey that was more than twice as long as one to the nearest land, and that would take at least two months. It was a terrible decision.

Adrift in the Vast Pacific

Within a week the men began to feel the desperate pangs of thirst. They had water with them, but were limited to only half a pint a day. Their boats also started to fall apart. The few Galapagos tortoises they had taken with them for food were soon eaten and their blood drained and drunk. In the unpredictable weather, and through the weakness of the men, it was also proving hard to keep the boats together. When they separated it took up too much time and energy to regroup. They soon resolved to continue alone if they separated again. By the seventeenth day they had covered over 1000 miles (1600 km), though their circuitous route still left them further from the coast of South America than when they started. They were, moreover, within a few days of Tahiti, and other islands of the Tuamoto Archipelago were even closer. Even in their appalling state, however, they would not be deflected from their intended course to South America, 2500 miles (4000 km) away. A fortnight later, having survived mostly off barnacles from the boats' bottoms, they had travelled a further 500 miles (800 km), but were still further away from South America than when they had started. Then they came across Henderson Island, which from the sea seemed a tropical paradise but proved to be a savage ridge of dry coral. At last, though, they found a spring and were able to build their strength for the next leg of the journey. Three elected to remain on the island and built a camp. The rest took to the boats. It was not long before Chase's boat separated from the others, which managed to stay together for over a fortnight more. By then the weakest had begun to die and they were swiftly butchered and eaten.

One of the boats was never seen again, but two were saved; one, containing five survivors, off the coast of Chile, 89 days after the *Essex* sank, and the other further south along that coast, 94 days after the wreck, and 12 days after the last man, Barzillai Ray, had died. The two survivors were found sucking the marrow out of his bones. Once aboard, fed and rested, their story poured out of them – disjointed, at times incoherent, and scarcely credible: the cause of all of their suffering had been the calculated vengeance of an enraged whale.

We are fortunate that one of the greatest authors in history was inspired by the tale of the *Essex*. So often are magnificent events recorded in dry lines, but through Melville we can glimpse those moments just prior to the strike, rendered vivid by a wordsmith of unparalleled genius in his descriptions of the sea:

'Suddenly the waters around them slowly swelled in broad circles; then quickly upheaved, as if sideways sliding from a submerged berg of ice, swiftly rising to the

surface. A low rumbling sound was heard; a subterraneous hum; and then all held their breaths; as bedraggled with trailing ropes, and harpoons, and lances, a vast form shot lengthwise, but obliquely from the sea. Shrouded in a thin drooping veil of mist, it hovered for a moment in the rainbowed air; and then fell swamping back into the deep. Crushed thirty feet upwards, the waters flashed for an instant like heaps of fountains, then brokenly sank in a shower of flakes, leaving the circling surface creamed like new milk around the marble trunk of the whale.'

Ever since, the stories of the *Essex* and of Captain Ahab's ship *Pequod* have been intertwined, and together they remain one of the most enduring tales of shipwreck ever told.

Iron, Steel and Steam

Nothing is more indicative of the impact that iron and steam had on the history of shipwrecks than the beautiful silhouette of the SS Great Britain *that graces the Bristol skyline today. This magnificent iron steamship now rests in the dry dock that Brunel had specially built for her construction; this is a ship that has retired to her cradle. By the very fact that she still exists, her history is intimately linked with the wider story of shipwrecks.*

Launched in 1843, the *Great Britain* was the largest ship yet built, and she was powered by a propeller, a technology that was only in its infancy and had never been used for a trans-Atlantic liner. She was also made of iron. Again, this was a new application of a new technology; some smaller iron ships had been built, but nothing on the scale of the *Great Britain*. She also had five watertight bulkheads. This was not a new invention, and had been common in

Chinese junks from the 13th century, but in European ships it certainly was new. Although there were still significant problems with the design and location of the bulkheads to ensure maximum efficiency in case they flooded, they nevertheless paved the way towards a new era of safety at sea in which damage to the hull could be isolated, and the immediate threat to the ship reduced until the time and resources could be found to make the necessary repairs. She even had iron lifeboats – the first merchant ship in history to have such equipment.

Iron for Peace and War

On 26 July, 1845, the *Great Britain* embarked on her maiden voyage to New York, a trip she completed in 14 days. Several further successful trans-Atlantic crossings ensued, but on 22 September, 1846 she ran aground at Dundrum Bay on the northeast coast of Ireland, the captain having mistaken the lighthouse there for that on the southernmost tip of the Isle of Man. She settled deep in the sand and gravel, astride a number of large rocks, and there she lay for over a year. For any other vessel built up to that time, this would have been exposure enough on a hostile shore to break her up, but the *Great Britain* withstood everything that the elements threw at her. The immensely strong iron frames that formed her hull were covered by a shell of iron plates, each double-riveted. Just above the keel, ten huge iron girders formed a platform inside the hull as a floor for cargo, and also provided a secure foundation for the boiler and engines. Combined, this gave her an internal strength that had never been matched in a ship. It was not long after her eventual salvage that she was at sea once again, but this time as an Australian emigrant ship, feeding the great Australian gold rush of 1851–2. She survived in various guises until, in 1886, in her last role as a clipper ship, a great fire in her hold forced her to turn back from Cape Horn for the Falkland Islands, and there she lay as a store-hulk until she was rescued in the 1970s and brought back to Bristol, with much of her original 1843 hull still intact. By then, she had therefore survived at least three events that would have destroyed any other ship: stranding, fire and abandonment in the windswept Falklands for 84 years.

The adoption of iron for shipbuilding also had very great implications for warfare in this period. Wooden warships could now be clad in iron armour, and iron warships themselves were soon built, first with the British HMS *Warrior* of 1860. These ships were almost indestructible if attacked with traditional artillery, and for a short period their loss through enemy action was highly unlikely. To combat this, however, far more powerful and more ingenious weapons were invented. Exploding shells, which had been used for some time for shore bombardment, were adapted for use in ship-to-ship engagements, and a great deal of thought was given to the ship's underwater vulnerability. Iron was extremely hard to repair in comparison to wood, and so even minor damage below the waterline could have grave consequences. The answer came in the form of the torpedo. Originally explosives attached to the end of long poles and detonated remotely, the locomotive torpedo – a torpedo that travelled under its own power – was invented by Robert Whitehead (1823–1905) in 1866. When united with the submarine, the torpedo could be launched in secret, and together they revolutionized warfare. In the coming century, torpedoes launched from submarines were to cause some of the worst shipwrecks in the history of the world.

Great Strides in Maritime Safety

While such leaps were made in the ability of men to sink ships, so too were advances made in the protection of life at sea from natural hazards. The increase in global shipping had inevitably seen an increase in deaths, and it became a political issue that could not be ignored. In 1867 alone 2340 British citizens perished in shipwrecks.

France, Scotland and America had demonstrated their commitment to maritime safety by centralizing control over lighthouses by the end of the 18th century, but it was not until 1836 that the numerous private lights that scattered the English and Welsh coastlines, many of which were poorly built, poorly positioned and poorly run, were all taken under the central control of Trinity House, which by then had begun to embrace its responsibility to build lighthouses in the most isolated of places. Treasured gifts to all mariners who frequent British coasts, in these years Trinity House established lights on the Bishop Rock (4 miles/6 km west of the Scillies), the Wolf Rock (8 miles/13 km off Land's End), and the Longships (1 mile/1.6 km west of Land's End). In 1886 the organization also replaced Smeaton's Tower on the Eddystone Rock off Plymouth. Similar American lighthouse building industry and endeavour are best summed up by the magnificent structure at Minot's Ledge in Massachusetts, at the mouth of Boston harbour. It is an outlying reef that has claimed as many wrecks as any other on the entire American coastline. The highest point of this rock was only 1 metre (3 ft) above the waterline at low water. Much like the Eddystone, early attempts failed but by 1855 an unprecedented masonry structure was begun. It took five years to build but still shines today.

All these towers, however, were built on rocks, and yet rocks are not the only underwater menace to mariners. Unmarked sandy shoals can cause a great deal of trouble and yet it was not until the late 1800s that the construction of a lighthouse on sand was even conceived. Hitherto, floating light-buoys and lightships had been used in limited numbers to indicate shoal water, and these had the advantage of being moveable to mark shoals that regularly shifted. However, they were always vulnerable to dragging their anchors or parting their cables. In most cases, light vessels and light-buoys were provided with a permanent support ship to help them regain their charted position after, or even during, periods of storm. If, indeed, their anchors did drag then their loss was almost guaranteed on the very navigational hazard that they were designed to illuminate.

The solution was to use a caisson, a large hollow tube sunk deep into the soft seabed, as the lighthouse's foundation. One of the greatest of these earliest lights was the Roter Sand light in the North Sea, roughly equidistant from the entrance to the Elbe river and the island of Heligoland. It was a critical development in German seafaring as it encouraged the expansion of the German trans-Atlantic liner service and the Imperial German Navy. The three major German ports of Bremerhaven, Wilhelmshaven and Hamburg are all totally dependent on the safety of this area of sea. The problem was solved by a prefabricated concrete caisson weighing 335 tons. When finally in place, the iron shell was over 30 metres (100 ft) tall, with as much as 13 metres (40 ft) of it buried in the sand. Nearly 5000 tons of concrete and stonework had been built into its elliptical centre. At roughly the same time a very similar lighthouse was built on the Fourteen-Foot Bank in Delaware Bay, on the eastern seaboard of the

United States. This was critical because the shoal obstructed the main shipping channel that ran towards the Delaware river, and from there to Philadelphia and its navy yard.

Battling the Elements

For those who were wrecked, cork life-jackets had been available from 1854, while in 1890 the first steam lifeboat was launched, showing herself capable of reaching over 9 knots in her first trials. Ten years later, the first steam lifeboat tragedy occurred when the *James Stevens*, one of the very earliest such vessels, capsized just offshore from her station in Padstow, on the north coast of Cornwall. Eight of her crew died, including her four engineers. The Royal National Lifeboat Institution (RNLI) immediately stepped in and granted £1000 for the immediate relief of their families.

Also, from the 1850s there was a rapid growth in the bureaucratization of seafaring. Ships that were unable to manoeuvre easily – such as telegraph-cable laying ships, or fishing boats carrying heavy nets – were given right of way. Sailing ships were to carry lights to identify themselves, as were steamships. It therefore became possible to gauge the capabilities of another ship at night, and therefore reduce the likelihood of collision. Rules for navigation in fog were also introduced. There were now legal requirements for the regular inspection of passenger liners, and there were standards of health and safety that had to be met. Distress signals were decided upon and international agreements drawn up.

On the largest and best-equipped vessels from 1881, electricity powered deck-mounted searchlights. Originally intended to illuminate torpedo boats, which made their audacious attacks under cover of night, they were also used as lights for search and rescue. Closely linked with the introduction of electricity was the science of weather forecasting, as by means of underwater telegraphic cables, news of storms could travel faster than the winds themselves. By the last quarter of the 19th century it became possible to track atmospheric changes across large portions of the globe. The introduction of the Beaufort Scale in the early years of the 19th century had also made it possible to describe weather and sea conditions accurately and uniformly. Together, it now became possible to monitor large portions of the globe's weather, and as more data was collected concerning weather history, so our understanding of weather systems grew, along with our ability to predict it.

Pioneering Meteorologists

In 1854 the precursor of the British Meteorological Office was formed, and was charged with collecting data. Its first director was Robert Fitzroy (1805–65), a man who did more to extend the science of weather forecasting in the hope of saving lives at sea than any other; he even invented the term 'forecast'. He compiled charts and tables of winds and currents worldwide to allow mariners to find the fastest and safest routes to their destinations. He invented 'synoptic' charts which showed the weather conditions in a given area at any one time, and in 1861 he sent the first storm warning to the Admiralty, Lloyd's and various ports. A few months later, he began to publish weather warnings in British newspapers. He was one of only a handful of men who understood that a falling barometer was a sign of forthcoming stormy weather and provided a number of annotated barometers in ports and harbours around Britain. His revolutionary idea was to save lives by preventing mariners from venturing to sea when poor weather was imminent.

Although he died by his own hand, depression clouding his once fertile and energetic mind, his endeavours have since been acknowledged and in 2002 the sea area Finisterre was renamed Fitzroy. Fitzroy had done so much, and yet the British government took a very dim view of this scientific 'prediction', and after his death they suspended research into weather forecasting. Meanwhile huge strides in weather forecasting had been made in America under the careful guidance of Joseph Henry (1797–1878) at the Smithsonian Institution and Matthew Maury (1806–73), an oceanographer. American forecasters benefited from the large size of their continent; they were simply able to collect much more data. By the 1870s, Austria-Hungary, Belgium, Great Britain, France, the German States of Prussia, Bavaria and Hamburg, Italy, the Netherlands, Russia, Scotland and the United States all had state-run meteorological services and institutions of one form or another.

The Plimsoll Line

Another significant invention of these years was the Plimsoll line, which brought the medieval practice of marking the safe load-line on a vessel up to date. Before then, British ship owners were allowed to cram as much cargo aboard their vessels as they could, at the expense of stability. It is well known that some even threatened the security of their vessels on purpose to guarantee that their ships would go down. Insured well above their actual value, the money was then claimed, and the shipbuilder turned a tidy, but brutal, profit. Such ships were known as 'coffin ships'. This was all put to an end by Samuel Plimsoll (1824–98), a politician with a deep concern for labourers in dangerous trades. He took the fate of shipwrecked mariners to heart and, in the teeth of fierce opposition from influential shipowners, championed the cause of improved safety at sea. Through his continual and impassioned lobbying of Parliament, the Merchant Shipping Act of 1876 obliged owners to mark their ships with a safe load-line.

For many, however, Plimsoll's name does not conjure up maritime safety, but children's footwear. Few know that the original 'plimsole' (as it is commonly misspelt) was named in honour of Samuel Plimsoll, as it was made with a rubber sole and canvas top, and therefore should only be immersed in water up to a certain point, like a ship.

The *H.L. Hunley*

17 February 1864

The submariner's life has always been fraught with particular danger. Such are the extreme psychological and physical demands of undersea warfare that navies only induct a select band of volunteers into their submarine services. The brief career of one of the world's earliest underwater craft attests to this pioneering spirit in all those who served and died in her. On her third and final voyage, the H.L. Hunley *earned her place in history by sinking the ironclad* Housatonic. *It was the first successful submarine mission in history, but it was achieved at a terrible cost to the attacker.*

The *H.L. Hunley* must assume pride of place in any book on shipwrecks. Not only was she the first submarine to sink an enemy warship, doing so 50 years before the feat was repeated, but she was also wrecked in the process. Furthermore, she had already sunk, and been raised, twice in her short life as the young sailors of the Confederate navy struggled with the new technology of submarine warfare. Perhaps no other vessel in history has been the subject of so many disasters at sea in such a short time, nor has a single craft been the focus of so many salvage projects. She was raised for the third and final time in 2000, complete with the remains of her final crew. These were men of courage and endeavour; men who knew that the *Hunley* had already sunk twice and killed all but four men of two entire crews, but who eagerly volunteered for the honour of attacking one of the Union navy's ships that was blockading the port of Charleston in South Carolina and squeezing the life out of the Confederate war effort. Until 2000 we knew very little indeed about this remarkable chapter of the American Civil War (1861–5), but now that the vessel has been discovered, excavated and raised, we are able to piece together a great deal more.

The *Hunley* was not named after a famous admiral, sailor, or battle; nor was it given an abstract title encapsulating the virtues of a navy: this is no *Victory* or *Indefatigable*. Fittingly, in this era of unprecedented invention in the midst of the Industrial Revolution, the *Hunley* was named after Horace Lawson Hunley (1823–63). A New Orleans lawyer and the Deputy Collector of Customs, he was her principal financial backer and one of a handful of men with an unwavering belief in her potential. There is always a great deal of money to be made from war, and Horace Hunley was not a man to overlook an opportunity. Huge rewards had been offered to anyone who could sink a Union warship, and Hunley was determined to write his name in the history books.

New Technologies

In practice, however, the problem of sinking an enemy warship was particularly acute in the early 1860s, perhaps more so than ever before. Wooden warships clad in iron armour, and ships made entirely out of iron, had demonstrated their worth in battle; traditional non-explosive iron shot simply bounced off such a protective layer. It was clear that advances in ordnance and gun founding were required before these armoured ships could be destroyed in the traditional way; Palliser shot, the earliest form of armour-piercing round, which was designed to be fired from rifled, muzzle-loading guns and punch through the thick wrought-iron plating on the sides of ironclads, was not introduced until 1867.

One obvious interim solution was somehow to target the enemy hull below the waterline. The plating of most ironclads did not extend far below the waterline, making them particularly vulnerable there, and the impact of damage to any vessel below the waterline was significantly disproportionate to its effect above water. A leak was paradoxically easy to identify but hard to locate, and in every instance it immediately drew men from the guns to assist with the repair or to man the pumps. Panic could easily take hold, but even if it did not, knowledge that their ship was letting in water wormed its way into the minds of every man aboard.

Almost a century earlier it had been discovered that a charge of gunpowder could be exploded underwater, and since then, numerous attempts using various techniques had been made to sink a ship in this way. The challenge resolved into somehow attaching an explosive device to the hull of the enemy ship before retreating to a safe distance and detonating it. One solution was for small, fast surface craft to dart in under the cover of night, and ram a long boom into the enemy hull, at the end of which was the explosive device. The charge would then be remotely detonated from a line that ran from the charge to the attacking boat. There were distinct disadvantages with this method, however, not least the difficulty in getting sufficiently close to an enemy ship without being discovered and destroyed first. Even then, the attacker had to retreat to a safe distance before exploding the charge, still without being detected.

The ideal solution, of course, was to approach the enemy underwater. This was not a new idea, but it had yet to be developed sufficiently for voyages underwater to be reliable and safe. Steam power could not be used underwater because there was no escape for the fumes of the burning coal, and effective batteries had yet to be invented. None of the three major requirements for the modern submarine existed, therefore: there was no engine; no light other than from candles; and no means of recycling air. There was, however, sufficient understanding of marine architecture and materials to construct an iron vessel that could, in theory at least, submerge and refloat using tanks filled or emptied of water as required, and which could proceed under hand-power without capsizing. The science of the submariner was therefore in its infancy, but it offered intriguing possibilities to the bold, and as always with the exploration of alien environments, such as space or flight, it attracted a loyalty and commitment in its advocates that bordered on the fanatical.

Horace Hunley was one of those men. Committed to discovery through new technology and new materials, he heard of a project to build a submarine proposed by two marine engineers from New Orleans, and he poured money into the scheme. Her name had not as yet been settled upon, but all involved in the scheme were determined that she would sail under a letter of marque to the glory of the Confederate navy. In practice, however, there was very little glory in the initial voyages of the *Hunley* and only mixed fortune in her last.

The 'Peripatetic Coffin'

We know that she was made out of a converted steam boiler, but only future research will tell us if it came from a ship or a train. This cylinder was enlarged and tapered at each end until she was 12 metres (40 ft) long, 1.5 metres (4 ft) high and just over 1 metre (3.5 ft) wide. In this tube sat eight men, cramped around the manual crank that worked the propeller, unable to stand up or move forward or aft until the craft was evacuated in single file through two hatches, situated at the bow and stern. The only light came from a candle, which also served as a warning that the oxygen was running low; it would gutter in time for the men to surface and open the hatches before dying from asphyxiation. A mercury gauge indicated her depth and a compass her direction. Near these controls the pilot sat at her wheel, which turned the rudder.

The *Hunley* had been shipped to the heavily blockaded Confederate state of South Carolina, and it was here that the first two tragedies occurred, both as she underwent trials in the calm waters of Charleston harbour. On the first occasion her captain, Lieutenant John Payne, found himself trapped in the hatchway as she prepared to dive. By accident he trod on the lever that controlled the *Hunley*'s planes, and with the propeller already rotating, she started to dive with the hatches still open. Five men drowned as the *Hunley* sank like a stone in 12 metres (40 ft) of water. So small and so near to shore, she was relatively easy to salvage, and the *Hunley* was soon ready for another test. Horace Hunley himself was so enraged at the apparent incompetence that had driven his vessel to the seabed that he insisted on commanding her to demonstrate that operating her safely just required a little experience and knowledge of the vessel's peculiar characteristics. As promised, Hunley took her down, around the bay, and up again with no adverse effects, and he promised to do so again a few days later. This time there was a great crowd. Hunley boarded the vessel, dived, and was never seen alive again.

A little too keen to dive, he let too much water too quickly into her forward tanks and she plummeted to the bottom, burying her bows deep in the Carolina mud. Hunley and another sailor were temporarily safe with their heads up in the entrance hatches, but the craft filled with water beneath them. They did not die immediately like their colleagues, but rather died of asphyxiation as they used up the air in the hatchways, up to their necks in water. This time we know a little more about the recovery of the vessel, which happened soon after she sank, and we know how deeply affected the Confederate general Pierre Beauregard (1818–93) was when the hatches were opened. He recorded in his diary:

> *'The unfortunate men were contorted into all kids of horrible attitudes; some clutching candles, evidently endeavouring to open the man-holes; others lying on the bottom tightly grappled together, and the blackened faces of all presented the expression of their despair and agony.'*

She had earned her nickname of the 'Peripatetic Coffin'. In ardent admiration for the project's benefactor, the name of the submarine, which hitherto had not been settled upon, variously being described as the 'fish boat', the 'fish torpedo boat' and 'the porpoise', became entrenched as the *H.L. Hunley*.

A Tempting Target

The death of Hunley shocked everyone, and yet it was also quite clear that the vessel had sunk through human error, and that the little craft was quite capable – as had been demonstrated on numerous other occasions – of undertaking a short voyage in calm waters. Despite these deaths, therefore, the momentum behind the *Hunley* mission did not falter; it was now just a matter of choosing her target. Soon the Confederate gaze settled on the Federal sloop USS *Housatonic*, a powerful new steam warship that carried 11 of the Union navy's largest guns. She lay in the middle of the northerly approach to Charleston, formidable both in her physical and symbolic defiance of the Confederate blockade runners, and representative of the mighty industrial resources that the Union forces could access. Everyone agreed that she should be the *Hunley*'s target, and the world would know that she had been sunk by Southern ingenuity and courage.

On the evening of 17 February, 1864, the *Hunley* set off

on her fateful mission and shaped a course for the distinctive silhouette of the *Housatonic*, then lying at anchor on blockade duty. At 8.45 p.m. the ship's master saw what he thought was a plank in the water. But then it stopped and changed direction. The alarm was raised aboard the *Housatonic*, but the submarine was already too close to make use of the ship's main armament. Pistol and rifle fire rained down on the little craft, which ominously stopped and then started to go astern. The *Hunley* had aimed for the mizzen mast of the *Housatonic*. The mizzen mast was both an excellent point on which the submarine could align herself, and just aft of it was the ship's main powder magazine and store of gun cotton – a type of explosive. If these ignited, the charge of the *Hunley*'s device would be magnified more than a hundredfold. The sharp turn of the deadrise would also have channelled the explosion into the hull and in effect created a blind spot where the *Hunley* could not be targeted from deck. Seconds later, a catastrophic explosion tore a huge chunk out of the *Housatonic*'s stern quarter and she started to sink fast. Eyewitnesses said that the after part of her spar deck was entirely blown off.

News of the sinking of the *Housatonic* spread like fire, but in its wake was a heavy silence regarding the fate of the *Hunley*. Signals were reportedly exchanged with a nearby shore station soon after the attack, but she was never seen again. Some believed she had become stuck in the hole she had made in the hull of the *Housatonic*, and had gone down with her; others believed she had been damaged in her own explosion or by small arms fire from the *Housatonic*. Those who knew how badly the *Hunley* performed in anything other than a flat calm blamed the rising wind and sea conditions, and believed she had foundered on her return.

Meticulous Archaeology

Archaeologists knew that many of these questions could be answered and the final few minutes of one of the most significant events in maritime history recreated if only they could locate the *Hunley*. In 1995 they found her buried in mud, only 300 metres (1000 ft) from the wreck of the *Housatonic*. The excitement was extraordinary. Were the remains of the crew on board, or did they manage to escape? If they were there, how had they died – from drowning, asphyxiation or from injuries received during the attack? Were the hatches closed from the inside, or open? Was there any damage to the submarine, and if so, had that damage been caused by small arms fire, the sea, or an explosion, or perhaps even a collision? These were just some of the questions that archaeologists wanted to answer, but as the investigation developed over time even more mysteries emerged.

The raising and subsequent excavation of the *Hunley*

has been unprecedented in its use of modern science to learn as much from the wreck as we can, while at the same time ensuring that the submarine and all of the artefacts inside her are conserved for future generations to enjoy. She has been marketed, and quite rightly so, as a 'gift from history', and the remains have been treated with a mixture of honour and professional rigour that has set the benchmark for future projects. Indeed, a significant part of the excavation planning included a formal burial with full military honours for the crew of the *Hunley* in Magnolia Cemetery in Charleston, where the remains of the *Hunley*'s other two ill-fated crews have also been laid to rest. Tens of thousands witnessed the event, and today those headstones exude a potent aura: these men lived together, died together, and have been buried together, all in a cause that tore America apart.

There is however, one important distinction between these three memorials: the crew of the *Hunley*'s last voyage were not buried as disarticulated bones, but as men whose faces we know, and whose histories have been teased from the past with exquisite care. That process began with a three-dimensional laser scan of the submarine, which allowed archaeologists to create a virtual image of her accurate to within 2 millimetres. This was then used to plan the safest means to gain access to the hull. Once inside, the mud was scraped away section by section, and gradually the *Hunley* gave up her secrets. The first artefact found was the wooden bench where the crew sat to turn the crank that powered the propeller. On the surface of the bench sat two small buttons from an artillery uniform coat, probably used as a cushion. A leather strap was found draped over the crank. Nearby, human remains started to appear, complete with shoes. They even found a tiny lead pencil used by one of the crew. Indeed everything that the crew had with them, that was not perishable, was discovered, even the lantern with which they signalled to the coast moments before sinking. Chillingly, the remains of the men were each at their allotted station in the vessel: once again the *Hunley* 's crew had died at their seats.

Personal Stories

One remarkable find was the identification tag ('dog tag') of a Union soldier named Ezra Chamberlain. It was found in close association with a skull, and it is assumed that the crewman had been wearing it around his neck. But what was a Union soldier doing on a Confederate submarine? Had he defected? Was he a spy? Or was he a prisoner of war forced to man this notoriously ill-fated vessel?

The mystery deepened when forensic scientists proved that the skeleton with which the tag was associated was around 30 years old, whereas it was known that Chamberlain was only 24 at the time of the *Hunley*'s mission. The answer was found in the archives: Ezra Chamberlain, eldest of four children and born in Connecticut, served in the Union army as a carpenter and soldier. He died in the failed attack on Fort Wagner, Charleston Harbour on 11 July, 1863. During these wars it was common practice to loot the enemy dead for money, equipment or souvenirs, and it is most likely that Ezra Chamberlain's tag was taken as a grim memento of this fight by one of the Confederate soldiers who had won the day. Perhaps it was taken from Chamberlain by one of the *Hunley*'s crew members, or perhaps it was acquired by him shortly after. In either case, it is most likely that the man who wore Chamberlain's tag as the *Hunley* sank was C.F. Carlson, a corporal in the Confederate army, who can be placed near Fort Wagner shortly after Ezra Chamberlain's death.

Undoubtedly the most poignant find was associated with the remains of the submarine's commander, Lieutenant George E. Dixon. Long after his death, a story about George Dixon survived and became part of the folklore of the Civil War. It was said that Dixon's sweetheart, the beautiful Queenie Bennett from Mobile, Alabama, had given him a gold coin at the outbreak of the war to keep near him at all times for luck, and he kept it in his trouser pocket. Then, at the Battle of Shiloh, Dixon was shot in the leg in the heat of the action, and the gold coin stopped the bullet. One hundred and thirty-seven years later, in the scattered pile of bones, clothes and personal effects that were identified as belonging to George Dixon, archaeologists found a gold coin, a $20 gold piece, deeply indented and with traces of lead in the scar. One side had been sanded and inscribed by hand with the following words: *Shiloh, April 6th, 1862, My life preserver, G.E.D.*

In time, historians researched the personal lives of each of the crewmen while archaeologists recovered their remains and anthropologists recreated their faces from the signature shape of their skulls. The work is ongoing to determine exactly how and when the *Hunley* sank. The latest clue is that the forward conning tower hatch appears to have been open when the vessel sank. It is unclear, as yet, if the tower was damaged in the explosion and the hatch could not be shut, or if it had been opened on purpose, perhaps to signal to the shore or in a vain attempt to escape. It might even have opened itself after a century of corrosion on the sea floor. What is certain, however, is that the hull of the *Hunley* was not designed to withstand the pressure of the explosion that she created, and it is most likely that her crew were stunned by the blast that tore open the *Housatonic*'s powder magazine.

In the final analysis, the story of the *Hunley* is a story of men prepared to sacrifice their lives in the name of the secessionist South, but it is also the story of men committed to the exploration of alien environments at any cost. Much like flight, pioneering exploration of the world underwater could not be made without death or injury, and no one knew this more clearly than those who chose to be buried alive in a steel coffin; who took that craft underwater, and then sat, unable to move, while the water rose from their ankles to their chests, then to their mouths, noses and eyes. Much of the *Hunley*'s fame now rests on our knowledge of her final crew, but of all the men who died in her, it is fitting that Horace Hunley's name endures in association with the craft. He gave birth to the *Hunley* through his money, energy and enthusiasm, and he gave his life to her as she took her first, faltering steps.

The *Titanic*

14 April 1912

Prior to her launch, the liner Titanic *was vaunted as an 'unsinkable' floating palace. But unlike many of her contemporaries in the golden age of trans-Atlantic travel, she was not fated to have a long career. As she neared her destination on her maiden voyage, an iceberg off the Grand Banks of Newfoundland tore a huge gash in her hull. The ensuing disaster, in which more than 1500 people perished, has resonated down the years. The discovery of the wreck by a remote-controlled submersible in 1986 fuelled further interest in the tragedy of the* Titanic.

Anyone who has taken a modern cruise purporting to offer first-class accommodation, should consider this: The À La Carte restaurant of the *Titanic* was designed as a Louis XIV banqueting hall, with floor-to-ceiling panelling of French walnut ornamented with carved and gilded mouldings, silk curtains, embroidered pelmets and beautiful garlands of flowers in bas-relief in the ceiling plaster. The walls of the First Class Smoke Room were panelled in mahogany in a Georgian style, and inlaid with mother of pearl. Through an intricate and ornate revolving door was the 'Palm Court', an area of trellises and climbing plants designed to replicate a portico of a fine country house. There was even a cosy fire in the reading and writing rooms.

'Empress of the Ocean'

She had been conceived in 1907 by leading executives of the White Star Line, encouraged by J. Bruce Ismay (1862–1937), the managing director. For almost 70 years, trans-Atlantic travel had grown at an astonishing rate since Isambard Kingdom Brunel had built the SS *Great Britain* in the early 1840s, and propeller-driven ships of iron, and later of steel, had first carried passengers from Britain to America. After the introduction of the turbine engine in the late 1890s the passenger liner business became obsessed with speed. Liners now competed for the Blue Riband, awarded to the ship with the fastest trans-Atlantic passage. Yet the White Star Line regarded itself as above such squalid competition. It believed in luxury; in excellent service in elegant surroundings. In 1907 the Cunard Line's *Mauretania* dominated the news with her record-breaking speed, while the White Star

Line's *Titanic* was designed to steal her headlines with splendour. Her keel was laid at the Harland & Wolff shipyard in Belfast in March 1909. Three years and one month later, approaching the end of her maiden voyage, she sank with catastrophic loss of life. The exquisite workmanship of her interior, which once defined her existence, can now only be seen from a remote camera launched from a submarine.

We are fortunate in knowing a great deal about the loss of the *Titanic*. Immediately after her sinking, there were two extensive enquiries, one on either side of the Atlantic, which necessarily reflected the scale of the public interest in her sinking. There was no time for the disaster to fade in the mind of the witnesses; indeed, the American enquiry began the day after the survivors returned aboard the *Carpathia*, which had picked up those lucky few who had made it aboard the lifeboats. Bruce Ismay himself was the leading witness, and he was one of 86. Eleven days later the British enquiry began, headed by the attorney general. Both enquiries have been published online since the late 1990s and rank as one of the internet's leading historical treasures. The accessibility of the material, combined with the publication of many other eyewitness accounts have led to an explosion in professional and amateur research into the *Titanic*, and there are few areas of research left untouched. Rather, our knowledge of that night undergoes a continual process of evolution as more facts and new angles continually combine to create alternative narratives.

In addition to this treasure trove of written material, we also have the wreck itself, discovered in 1986. The *Titanic* sank in over 12,500 feet (2.4 miles/3.8 km) of water, but with the use of robotic vehicles with mechanical arms and a camera, it has been possible to explore the ship. We know from the discovery of the wreck that she broke in two before sinking, and her stern now lies a full 600 metres (1968 ft) away from her bow section, facing the opposite direction. With a gross registered tonnage of 46,328 tons, but with a deadweight of almost 70,000 tons, the *Titanic* plunged to the seafloor at such speed that the water inside the hull tore the decks apart, and when she finally crashed down in the mud, many of the artefacts inside her were forced out of the hull. Some of that material came to settle over half a mile (1 km) away, and the debris is scattered over a square mile of the sea floor.

A Night to Remember

At first glance, the story of the *Titanic* is quite simple. Bound for New York, with a passenger list that included some of the world's wealthiest and most prominent people alongside middle-class families and poor migrants, she left Ireland amidst a great fanfare, and carried the hopes and dreams of all those aboard across the western horizon.

Then, on the night of 14 April, she struck an iceberg and sank 2 hours and 40 minutes later. Records do not agree, but some 1500 people died when she slipped into the inky Atlantic on a starlit and eerily calm night. When one probes a little deeper, however, the *Titanic* story becomes multi-layered, elusive and contentious.

Perhaps most famously, there were insufficient lifeboats for her full complement of passengers and crew (2208). This resulted in a strict policy of 'women and children first' to man the lifeboats, which raised as many problems as it solved. First-class women and children were certainly favoured before the rest. Only one first-class child died, and she did so because her mother refused to leave her husband, and the child in turn refused to leave her mother. In contrast, almost all of the children from the poorest class of passenger, the steerage class, died. Indeed, one of the most shocking aspects of the *Titanic* sinking is how few of the accounts from first-class passengers mention the terrible loss of life they must have witnessed, and how many are concerned with the loss of jewellery, furs and pets. The startling lack of accounts from the poorest class of passengers (because so many of them died) is equally powerful.

Not only was there a preference system for the manning of the lifeboats, but some chose not to man the lifeboats at all. It was believed by some, quite wrongly, that the *Titanic* was unsinkable, an idea whipped up by the enthusiastic media before her launch, and to abandon oneself to an open boat in mid-Atlantic did not seem sensible to all, especially to many of the elderly, young or frail who needed the lifeboats the most. Others simply refused to leave loved ones. The first lifeboat, with a maximum capacity of 40, was launched with only 12 aboard. Lifeboats 6, 7 and 8, each with a full complement of 65, held only 28 each. In their defence, the officers who lowered the boats would not have known at this stage how much time they had left, and it was vital that the ship did not go down with the lifeboats still attached. Moreover, few passengers were willing to take the lead and abandon the luxury of the ship for the stark reality of the freezing North Atlantic. In some respects, then, the luxury of the *Titanic* itself exacerbated the death toll.

Nevertheless, it is because the story of the *Titanic* has such socially and politically charged undercurrents that it has endured for so long in the popular imagination. Many imagine it to be the worst maritime disaster in history, but it is not; in fact, the death toll of the *Titanic* is less than one-sixth of those who died aboard the German troopship *Wilhelm Gustloff* (see pages 268–73), torpedoed by a Russian submarine in 1945, killing more than 9000. The death toll of the *Titanic* is also less than the worst maritime disaster in American history: 200 more died when the Mississippi river steamboat the *Sultana* exploded in 1865, but neither of these tragedies so magically captured the essence of their age as did the *Titanic*.

One famous tale of stoic heroism on board the *Titanic* was that the ship's band did not abandon their instruments and take to the rafts but rather, recognizing the importance of music in maintaining an air of calm, continued playing even as the water lapped over the bow. So too are the engineers celebrated in the *Titanic* story. Many inboard areas of the ship had no natural light, and as it was night-time, the generators were crucial to supplying those areas with electricity. The engineers stayed at their posts in the bowels of the sinking and listing ship and kept the generators running. It is why every image one sees of the *Titanic* sinking shows her lit up like a Christmas tree. By supplying light, they too played their part in averting panic, and also helped guide the *Carpathia* to the wreck site. Monuments dedicated to the memory of the *Titanic*'s engineers and musicians were soon erected in Southampton; more than one-third of all those killed on board – 549 – were crew members from her home port. The memorials can still be seen in the city's parks. That of the engineers is the most striking. It is 5.7 metres (19 ft) high by 9.7 metres (32 ft) long, with a 2-metre (7-ft) tall figure of 'angel glory' holding a laurel wreath in each hand. The engineers are depicted at their posts in bronze bas-relief beneath the angel. It is a monument to duty, courage and stoicism, but above all to steadfast 'Britishness'.

In a similar vein, those white, Anglo-Saxon voices also cherished stories of foreigners and the lower classes challenging authority at the very moment when the need to maintain discipline was paramount. On the other hand, for those whose sympathies lay with the poor, to build a first-class promenade deck uncluttered with lifeboats, so as to increase the space for first-class passengers, was a capital crime in itself – as it was for the crew to spend the night rearranging the deckchairs when they should have been looking out for icebergs. These voices chose to believe the tales that the first-class men were only held back from the lifeboats at gunpoint.

Separating Fact from Fiction

Thus pressed into service for political ends, the facts themselves have become greatly confused from the moment of her destruction. Hollywood director James Cameron's 1997 film *Titanic* has further merged fact and fiction, and has introduced whole new distortions to the tale, as it deliberately brings ideals precious to a 1990s, and largely female, audience to the centre of the story. One should not underestimate the power of this film in forming people's views of the *Titanic* disaster; a budget of $200 million buys a lot of collective memory.

All of this has to be carefully unpeeled to get as close to the true story as possible. The most recent research into the fate of the third-class passengers, more than half of her entire passenger list, has refuted the popular story that they challenged the authority of the crew as the ship sank. Nor is there any evidence that large numbers of them were forcibly kept below decks, as depicted in the 1997 film. The only evidence of anything like that happening comes from an isolated incident between two people which is uncorroborated by any other source. Likewise, there is no evidence that any officer shot any of the passengers or committed suicide, both of which featured prominently in Cameron's film.

The *Titanic*'s demise at least had a number of important ramifications for maritime safety, and led directly to the first International Convention for the Safety of Life at Sea in London on 12 November, 1913. Both enquiries had concluded that a contributory factor to her sinking was that she was travelling too quickly in a sea area known to be littered with icebergs. Her captain, Edward Smith (1850–1912), had received numerous iceberg warnings and yet, under pressure to make the crossing to New York in only six days, had not slowed his vessel. Smith claimed that it was normal practice to maintain high speed even after iceberg warnings; the assumption being that an iceberg large enough to sink a ship would be seen in sufficient time to be avoided. After the enquiry, the British Board of Trade required all vessels to reduce speed in hazardous conditions.

A major factor in the large loss of life was the inadequate number of lifeboats, although she had been fitted with enough – 16 wooden and four canvas collapsible boats – to satisfy the legal requirements of the day. But as the tonnage of ships had grown exponentially since the early 1900s, the legal requirements were long outdated. After the enquiry it became mandatory for every ship to carry sufficient lifeboats for all aboard. Yet the loss of life was not solely due to the lack of lifeboats. The initial distress flares sent up by the *Titanic* had in fact been seen by another liner nearby, the *Californian*, but her radio operator had long since gone to bed and turned off his equipment. Unable to hear the distress call, it was assumed that the *Titanic* was having a party with fireworks. After the London conference, it became law that ships' radios were manned 24 hours a day.

These, then, are the most significant strands of the rich story of the *Titanic*, yet it is ironic that one of the most significant factors in her sinking has consistently been overlooked. As modern science reveals the mysteries of our environment, it becomes increasingly clear that the wreck of the *Titanic* was not only a product of its age because she was especially luxurious, or large, or because there were insufficient lifeboats, or because many more wealthy passengers were saved than poor, but because she struck an iceberg. Ever since the disaster, the International Ice Patrol has monitored icebergs that drift down the Labrador Coast from western Greenland; in an average year there are 500. But in 1999, for the first time in 85 years, the shipping lanes southeast of Newfoundland were completely free of icebergs. Indeed, since the 1980s winter temperatures to the north of the Grand Banks have risen by 0.5°, enough to change forever the landscape in which the *Titanic* sank.

The *Endurance*

21 November 1915

The late 19th and early 20th centuries were characterized by an insatiable appetite for exploration, as the major European powers vied for military and colonial supremacy. In this period, one expedition stands out for the exemplary courage and fortitude of its members – Ernest Shackleton's Imperial Trans-Antarctic Expedition of 1914–16. Against all the odds, he and his crew survived appalling conditions after their ship was lost, all returning home safely. In particular, the epic voyage in a small boat by six of the party to ensure the rescue of their shipmates from a remote island remains one of the most remarkable feats of seamanship ever undertaken.

On 30 december, 1913, the polar explorer Ernest Shackleton (1874–1922) was photographed on a London street. It was the day after he had announced his latest expedition to the Antarctic, and this time he would cross it on foot. The years leading up to the First World War were a time of heightened nationalism. British maritime superiority, and hence her security as an island nation, was being openly challenged by Kaiser Wilhelm II of Germany (1859–1941), whose formidable navy loomed on the eastern horizon. Anglo-German rivalry was also fierce in Antarctic exploration. The Antarctic was one of the few remaining areas of the world yet to be conquered by man; the South Pole had been reached in 1911 by the Norwegian explorer Roald Amundsen (1872–1928), but now a number of Europeans were desperate to be the first to cross the continent, and Shackleton's main rival was the German, Wilhelm Filchner (1877–1957).

The British, therefore, had every reason to be proud of their Antarctic explorers; although a sideshow to the impending war, polar expeditions provided Great Britain with an opportunity to demonstrate her superiority over her continental rivals. The announcement of Shackleton's latest exploration was headline news, and this seemingly unremarkable photograph of him made it onto the front page of the *Daily Mirror*. But on closer inspection the photograph was really quite remarkable: 30 December, 1913 was one of the coldest days in London that anyone could remember, and Ernest Shackleton is the only one not wearing a heavy overcoat or gloves. The *Mirror* rejoiced – here indeed was the essence of British explorers! Here was a man whose distant gaze was fixed on a desolate land on the far side of the world. Here was a man who would do what it took to fly the British flag where it had never been flown before. A man who stood out from the crowd; a man of courage and determination. Above all, here was a man of endurance.

Aptly Named

This was the age in which stirring adventure novels such as Henry Rider Haggard's *King Solomon's Mines* (1885), H.G. Wells' *The First Men in the Moon* (1895) and Arthur Conan Doyle's *The Lost World* (1912) were all the rage, and Shackleton could have been a character in any of them. 'By Endurance We Conquer' (*Fortitudine Vincimus*) was even his family motto. Perhaps the need for endurance in life had been drummed into him from an early age, for it was also in his nature, and when he came to name the ship that would take him and his men to the edge of the known world, in the most hostile climate imaginable, Shackleton did not waver. She would be called the *Endurance*. Except for Nelson's *Victory*, no ship has ever been more fittingly named for the conduct of her commander and crew.

The *Endurance* was originally named the *Polaris*, and was built in the deep fjords of Norway specifically for the ice. Her hull was narrowly tapered so that she might be squeezed upwards and out of the ice should she become trapped. It was also exceptionally strong. She had been built for two Norwegian entrepreneurs, Lars Christensen and Adrien de Gerlache, who had envisaged 'polar safaris', but after she was built, the scheme collapsed and she lay idle. When Shackleton discovered her he snapped her up.

After years of preparation, the expedition was finally readied, but war drew ever closer. With the imminent prospect of conscription, Shackleton offered every man in his expedition to the Admiralty. Winston Churchill (1874–1965), then First Lord of the Admiralty, refused and ordered Shackleton to the Antarctic. The *Endurance* left South Georgia on 5 December, 1914. Her reinforced hull allowed her to nose through the ice pack until she was within 80 miles (128 km) of land, but by 18 January, 1915, the floes caused by a particularly cold winter had closed around her and held her fast. The ship's captain, Frank Worsley (1872–1943), told Shackleton to prepare for the worst. '*What the ice gets, the ice keeps*,' he said.

For ten agonizingly slow months she was crushed in a vice of ice, forced up and out of the sea in an unnatural dry-dock while her ribs and back were bent, lifted and slowly broken. Her captain and crew looked on helplessly. The *Endurance* was their home, their warmth, their transport and, as she was crushed, so were their hopes. Finally, on 21 November, 1915 – a month after being abandoned – she could take no more and sank through a hole in the ice, leaving the crew adrift on a floe in the Weddell Sea. Their last link to civilization was broken, but in this defining moment Shackleton's character stood out. He did not bemoan his misfortune, he did not regret the past; he simply looked to the future, a future of his choosing. '*Ships and stores have gone*,' he wrote, '*so now we'll go home*.' What followed is one of the greatest ever tales of endurance and without doubt the greatest adventure story of the 20th century. In fact, they took with them a great deal of stores; the ship had, after all, been equipped for a polar land expedition. They also saved her three lifeboats, which could take all of the men.

An Agonizing Ordeal

The first major obstacle was to get near the sea, for by now the ice that had trapped the *Endurance* had solidified. They loaded the boats with stores until they weighed as much as a ton each, and then dragged them over ridges, furrows and crevasses. Less than a mile and a half (2.4 km) from the wreck they admitted defeat and set up camp to wait for the ice to thaw, and gradually it did. Their camp rotated as the ice shifted and they drifted northwards at roughly two miles (3 km) a day. Once the weather improved they set off for a second time, covering little over a mile and a half in every eight hours of marching. They marched for eight miles (13 km) until the floes became unstable, but then were unable to retreat.

In the following days the sled dogs were shot; this makes for some of the hardest reading of all of the diary entries. '*Hail to thee old leader Shakespeare,*' wrote the expedition's photographer, the Australian Frank Hurley, of his favourite dog, '*I shall ever remember thee – fearless, faithful and diligent*'. Still the ice did not break, however, although by now they had drifted within 150 miles (240 km) of a known, if uninhabited, island. A full four months later the floe broke and the men prepared the boats. Once at sea they rowed clear of the ice, but all the time drifting away from dry land. Eventually they made it to open sea and a helpful wind blew them towards Elephant Island. Still ten miles (16 km) away, exposure in the open sea in freezing temperatures had badly affected all of the men. A number collapsed with fatigue, others with an overpowering sense of helplessness and hopelessness. Some had not slept for more than 90 hours and no one had enjoyed hot food or drink in that time. When they finally landed they had spent seven days in open boats in the South Atlantic, but tellingly almost all of them were too happy to sleep that night. None of them had set foot on dry land since leaving South Georgia in early December 1914. It was now 15 April, 1916.

Still their ordeal was not over, however. There was a ready supply of seals, birds, seaweed and an abundance of fresh water, but Elephant Island was nearly 1000 miles (1600 km) from any form of civilization, and nowhere near any regular shipping lane or whaling ground. On 24 April, six of the men, led by Shackleton, took the 22-foot whaleboat *James Caird* across the Southern Ocean, one of the coldest and stormiest seas in the world, to land in South Georgia 16 days later. Shackleton left his most trusted colleague and friend, Frank Wild (1873–1939), in charge of the shore party. The letter Shackleton left with him survives. An excerpt from it reads:

> *'In the event of my not surviving the boat journey to South Georgia you will do your best for the rescue of the party…I have every confidence in you and always have had, may God prosper your work and your life. You can convey my love to my people and say I tried my best.'*

And with that, Shackleton and his small crew left to try their best, which was better than any could have believed possible.

Shackleton described that open-boat journey as one of '*supreme strife*'. It was dark for 13 hours of every day, and they witnessed a storm unlike any they had seen before. In the worst of the weather they were all soaked every three to four minutes. The clouds and rain hid the sun, moon, stars and horizon and yet still Frank Worsley navigated them with unerring accuracy towards their destination. Ice soon clad the little boat. Three times the hull, the deck and the sails all had to be cleared of ice with knife and axe without damaging their precious transport. When they finally struck land they were 150 miles (240 km) from the nearest whaling station, but none had tasted fresh water for 48 hours and all were near physical and mental collapse.

They had landed on the opposite side of South Georgia to the whaling station, and in between lay an unclimbed and unsurveyed range of mountains. There was little choice in how to proceed; the *James Caird* could take them no further and winter was rapidly closing in. Days were getting shorter and the weather worsening. Their only option was to walk to safety. Shackleton took with him the two toughest and most durable of the party, Frank Worsley and Tom Crean, and they set off with a length of rope and a carpenter's adze. Their feet were frostbitten and they had taken no real exercise for six weeks. They rested for one minute every quarter of an hour. Twice they climbed icy peaks, cutting steps into the ice for every step, and twice they found no route through. On their third attempt they had a breakthrough, but the start of the descent was desperately steep, night was coming on and they were still over 1500 metres (5000 ft) above sea level. To cut steps and climb down safely was taking too long; their only option was to slide. The three men sat one behind another on their coil of rope and pushed off. They slid, whooping and cheering, almost 500 metres (1500 ft). The rest of their journey was desperately difficult but never as dangerous, and they made it to Stromness Bay 36 hours after setting out. Soon after, rescue missions were launched to pick up the men on the other side of South Georgia and those still ashore on Elephant Island. Every man in Shackleton's original party survived.

Historians studying the wreck of the *Endurance* and her crew's fight for survival are fortunate in that the expedition is extraordinarily well documented. When the men first left the site of the shipwreck they had to lighten their loads as much as possible. All personal items were to be discarded, and as ever, Shackleton both knew the way and showed the way, tossing his gold cigarette case into the snow at his men's feet. His Bible and a few gold coins followed. The men followed suit and threw away all their personal possessions, with three exceptions: the banjo was to go with them to help morale, Frank Hurley was allowed to keep his camera, and every man was allowed to keep his personal diary. Because those diaries were saved, and so many carefully stored upon their return, we can now follow the expedition day by day from the perspective of a variety of characters. Here is the ever-optimistic Able Seaman Timothy McCarthy as the little *James Caird* fought the might of the South Atlantic, ice weighing the boat down, and snow blizzards obscuring the sun: '*It's a grand day, sir!*' he said, grinning at Worsley. And here, in surgeon Alexander Macklin's diary on the first day at Elephant Island, is his bare desperation at being unable to make their shelter, an upturned boat, waterproof: '*I think I spent this morning the most unhappy hour of my life – all attempts seemed so hopeless, and Fate seemed absolutely determined to thwart us. Men sat and cursed, not loudly but with an intenseness that shewed their hatred of this island ...'* Finally, there cannot be many diary entries as powerful as that of Shackleton as the men drifted on an ice floe in the desolate wastes of the South Atlantic, desperate for a glimpse of land: '*64°. 16 ½' S., 52° .4' W. No news; S.E. wind, fine weather. Patience. Patience. Patience.*'

A Very British Hero

One of the most striking things about these diaries is the influence of Shackleton on everyday events. On previous polar expeditions, infighting and incompetence had led to scurvy, fighting, maiming, death, murder and suicide. Charles Hall (1821–71), for example, leader of the American attempt to reach the North Pole in 1871, was poisoned by his own men. And yet from the diaries of Shackleton's men we have tales of endless competitions, challenges and great games of poker in which luxurious imaginary bets of tickets to the theatre, or silk scarves, were made, and recorded, to be repaid on their return. The writing of poetry became almost competitive and the men queued up for accompaniment from Hussey's banjo. If one image sums this up more than any other, it is of Shackleton's crew playing football on the ice with their stricken ship in the background. In 1964 the surviving members of the Shackleton party met in London for the 50th anniversary of their departure from the Antarctic. With half a lifetime to consider the ways and means of their survival, the *Endurance*'s first officer, Lionel Greenstreet (1889–1979), was asked why they had survived when so many others had died. He answered in one word: '*Shackleton*'.

In the most unforgiving analysis, Shackleton's expedition was a failure. He did not do what he set out to achieve; he did not cross the Antarctic. On the other hand, although Robert Falcon Scott (1868–1912) did reach the South Pole, he died on his return, and yet for years enjoyed a far greater reputation. Recently, however, as Shackleton's journey has been retold in print and on screen, he has become cherished as a typically British hero: a hero who made a great success out of failure. In many ways, then, Shackleton's story resembles that of the evacuation of Dunkirk in 1940: a tale of triumph over adversity, of failing with spirit. These are themes that are dear to the British and which permeate a number of great shipwreck stories: everything can go wrong but you can still triumph.

World Wars

The 20th century dawned with two inventions that would revolutionize seafaring. In the early summer of 1895, Guglielmo Marconi, 21 years old, transmitted a radio signal a few metres between two insulated plates separated by a spark gap. By August 1895 he could transmit a signal 2 miles (3 km). He took his invention to Britain, and by 1897 had established radio communication across the Bristol Channel. Not long afterwards, Queen Victoria, then at Osborne House on the Isle of Wight, communicated with the Prince of Wales on board the Royal Yacht in the Solent. In 1898 a signal was sent across the English Channel. Just three years later a radio signal crossed the Atlantic.

The implications were huge. Ships could now communicate with each other and with the shore over vast distances. It was now no longer necessary to carry a watch at sea as the time was regularly broadcast from Greenwich.

If a ship carried a direction-finding aerial, then it could pinpoint its location by triangulation from two or more radio signals broadcast from known positions. Weather warnings could also be given, and distress signals made and received. The very first one was made on 23 January, 1909, when the British White Star liner *Republic* outbound from New York to the Mediterranean, collided in thick fog with the *Florida*, an Italian liner bound for New York. Jack Binns, radio operator of the *Republic*, sent the first CQD message – the precursor of the SOS. The *Republic*'s sister ship, the *Baltic*, came to her rescue. By broadcasting that message, Binns saved the lives of everyone aboard both liners – 1290 souls.

The second significant innovation was the gyroscopic compass. Traditional compasses were magnetic and thus pointed to magnetic north. It had been known for centuries that local magnetic fields affected a magnetic compass, and this became increasingly problematic when ships were no longer made of wood, but of iron and steel. It became worse after periods of intense activity aboard a warship, such as gunnery practice, which further disrupted the ship's magnetism. The use of the magnetic compass was also difficult aboard submarines, as the latest designs completely shielded the ship's instruments from the earth's magnetic field when submerged. They were also crammed with electrical equipment and motors that affected a magnetic compass when it could be used on, or near, the surface. A gyroscopic compass, however, does not point to magnetic north, but uses the relationship between its own revolution and that of the globe to orient itself to true north. By 1912 one that could cope with the roll of a ship had been produced, and the art of navigation finally caught up with the new materials of shipbuilding.

Increased Danger of Shipwreck

A recurring theme in the history of shipwreck is that every leap made to improve human safety at sea is matched by the invention of new ways of imperilling it further. Thus the use of iron and then steel in ship construction made them less susceptible to fire and damage from grounding on rocks or reefs, but at the same time the perfection of torpedo design made it far easier to sink a ship in an act of hostility than ever before. Moreover, one of the lessons of the *Titanic* disaster is that the introduction of new technology is not an end in itself; it must first be understood and then deeply integrated into the practices of the time. Thus Marconi's invention of the radio transformed people's ability to protect themselves from shipwreck, but it had not been sufficiently acknowledged by 1912 for the *Titanic*'s radio set to fulfil that role; her Mayday messages went unheeded. Now with global war fast approaching, it would not be long before devices were invented specifically to jam or to scramble radio transmissions. The potential to improve the protection of life at sea is already threatened, therefore, but in times of war it becomes more difficult still. Indeed, in the 31 years between 1914 and 1945, the likelihood of shipwreck was higher than ever before.

Part of the reason came from the sheer number of vessels at sea in these years. Both world wars were characterized by the significance of maritime trade that generated the wealth to pay for the wars, and then sustained them by keeping the soldiers, sailors and airmen fed, clothed and armed. Meanwhile, food convoys kept those on the home front alive, which in turn gave the troops another reason to fight. Vast convoys of merchantmen plied their trade, protected by naval escorts, and hunted by packs of submarines. The navies themselves were larger than ever before; at the Battle of Jutland in 1916, 250 ships faced each other across the North Sea. Invasion flotillas of unprecedented numbers gathered together to carry tens of thousands of men, supplies and transport across small stretches of water and open-ocean passages of hundreds, and occasionally, thousands, of miles. These were only matched in scale by evacuation fleets, as troops, women and children fled from rampaging armies.

New Threats – Submarines and Aircraft

It was inevitable, therefore, that shipwreck in these years would rise through accident alone. But at the same time, unprecedented weaponry ravaged those fleets. By the early years of the First World War, the submarine had become an operational reality. They could stay submerged for a matter of hours, running on electric motors or diesel engines, and the torpedoes they fired were fast and swam in straight lines. At the same time, there were precious few weapons that could be used against these submarines until efficient depth charges were developed in the Second World War. In the first war, primitive bombs were dropped on submarines from airships or aircraft, usually with little effect. If a ship's captain was fortunate enough to catch a submarine on the surface they could be rammed or targeted with machine-gun fire. Ultimately, however, it was a permanent presence of naval ships around the convoy vessels, many of which had been armed themselves, that temporarily countered the submarine threat; they could not fire without revealing their position, and if surrounded with sufficient firepower, these lightly constructed, slow and otherwise fairly primitive submarines would be lucky to escape. Escorts also used the hydrophone to 'listen' underwater, which helped identify a submarine threat.

Much the same pattern continued in the Second World War, but the submarines could stay submerged for longer, they could go deeper and they could travel faster. Anti-submarine warfare at the same time rested heavily on the initial knowledge of a submarine's location. Then it could be targeted with depth-charges, or hunted until it was forced to re-surface. Again, air-reconnaissance had a significant part to play, but more sophisticated sonar technology allowed ships to 'see' underwater with sufficient clarity to locate a submarine. Meanwhile, land-based cryptographers deciphered radio intercepts that helped surface ships locate their prey.

The submarine was not the only new threat in these years. After the First World War, a great deal of energy was directed towards the ability to launch aircraft from ships, and then to recover them. Sea-planes that were catapulted from the parent ship or dropped into the sea made way for flight decks from which aircraft could be launched in a much more traditional style. By 1917 this was possible, if the science of it was still a little unpolished. By the mid-1920s aircraft carriers, much as we would recognize them now, with a flight deck the full length of the hull, had been built. In the coming years, ships would be targeted by both bombs and torpedoes launched from aircraft, and even, in 1945, by aircraft themselves. In response to unrelenting pressure on their manpower and military resources from American naval success, the Japanese began to use *kamikaze* planes to target ships. These were aircraft loaded with high explosives that were flown directly into their targets. Likewise, single-man, guided suicide torpedoes known as *Kaiten* were launched from Japanese submarines in the final years of the Second World War.

Terrors facing the Wartime Sailor

These turbulent years, therefore, were characterized by a profound change in the experience of the common sailor at war. No longer was the enemy a visible focal point for anger and aggression, in which proximity was the ultimate demonstration of aggressive intent. Now, the enemy had become elusive, only a metaphorical 'focal' point for those energies. With submarine, airpower and long-range guns that could accurately target another ship well beyond the horizon, the enemy had become invisible. No longer was it possible to draw a measure of confidence from the look-out's silence. The power of the human eye as a defence against shipwreck had become drastically reduced, and so had the sailor's sense of security. These were years in which the psychological challenge of seafaring gained a new dimension. Men lived in continual fear of their lives from an unknown quarter. Some were paralysed by fear, while others were energized into action, but all expected to be woken at night by an unheralded and catastrophic explosion.

Nor was this the end of the wartime mariner's concerns, however, which are best summed up by the tragedy of the *Rohilla*. She was an 8000-ton ship of the British India Line that had been converted into a hospital ship for the First World War. Based in the Firth of Forth, she was ordered to run to Dunkirk. Clearly marked as a hospital ship, and therefore not a legal target for enemy attack, she nevertheless had to run the gauntlet of her own coast. Just across the North Sea lay the mighty German battle fleet, and for many months the British government feared a maritime-borne attack on the east coast. To make it as difficult as possible to execute such an attack, every coastal light on the entire east coast of Britain was extinguished. As the *Rohilla* made her way to northern France in October 1914 a great gale struck. Unable to ascertain exactly where she was, the *Rohilla* was driven towards the terrible rocks off Whitby known as the Saltwick Nab. The men and women of the *Rohilla* were eventually rescued, including a lady named Mary Roberts (1881–1933) who had also survived the wreck of the *Titanic*, but the resources of the Whitby, Scarborough, Teesmouth and Tynemouth lifeboats were all stretched to the limit in doing so.

Courageous Volunteers

Three of the four attempts to rescue the passengers of the *Rohilla*, moreover, were carried out under sail and oar, much as they were elsewhere; engines of any description had not yet become standard for lifeboats. Indeed, it was not until 1948 that the last sailing lifeboat was withdrawn from service. There was also a nationwide shortage of skilled and fit lifeboatmen, since all of those who were sufficiently able and young had been drafted into the military. By 1918, for example, there were only 18 members of the Lowestoft lifeboat, and all were over 50 years old. Twelve of those were over 60 and two were over 70. Nevertheless, the elderly men of the Lowestoft lifeboat were still hardy and capable, and all had a deep knowledge of the sea. They covered themselves in glory more than once. With such limited resources, British lifeboats in 1916 rescued 1185 lives, hitherto the largest number in the institution's history. In the next war, British lifeboats put to sea 3760 times, 2212 of those to ships in distress from enemy action.

Equally courageous were those elderly men who manned the few lighthouses and light-vessels that were kept burning during the war. The validity of a lighthouse as a military target has always been rather ambiguous. The most significant lights are useful to all shipping. Local knowledge of sandbanks, reefs and rocks is simply negated at night; British lights in the western approaches, for example, were as important to British fishermen from the Isles of Scilly, bred in those waters, as they were to German U-boat captains from Wilhelmshaven who had never before ventured out of the North Sea. When the first Eddystone lighthouse was under construction in 1697, French privateers, at war with England, landed on that isolated rock and captured the unfortunate masons and the lighthouse designer, Henry Winstanley (1644–1703). When news of their capture reached King Louis XIV he exploded with rage. The crew of the French privateer were immediately imprisoned, the British masons released with a royal pardon, and returned to their task. '*Their work is for the benefit of all nations*,' Louis is reported to have remarked; '*I am at war with England, not with humanity*'.

In many circles this sentiment endured, but in others the strategic significance of a light prevailed. In 1939 one of the earliest targets of the German *Luftwaffe* was the undefended lightship at East Dudgeon. She was sunk by bombs, but some of her crew survived to row ashore to the Norfolk coast. Having rowed all night in ferocious weather the little boat was wrecked and all but one man died. Not long afterwards a machine-gun attack on a Humber light-vessel killed eight of the nine-man crew. Numerous attacks on lighthouses followed, many of which had been converted by the army for auxiliary use. The lighthouse at Ouistreham in Normandy, for example, was used as an artillery observation post by British forces in 1944, and the lighthouses on the Italian island of Gorgona housed a hydrophone-listening point.

International Conventions

In spite of all of these setbacks to the challenge of safeguarding life at sea, great advances were made in the first half of the 20th century. Although many of the rules and regulations passed before the wars were temporarily suspended by hostilities, their very existence did much to accelerate international agreement as peace settled once more on most of the world's oceans, and nations mourned their dead. Indeed, a pattern of growing international agreement regarding safety at sea can be traced back to the late 19th century, which was then energized by the loss of the *Titanic*. Hitherto, territorial waters were governed by each nation's widely diverse (or sometimes entirely non-existent) regulations concerning safety of every type – from load-lines and construction techniques to rules for navigation in a fog and right of way. There were very few agreed rules at all for international waters.

After the loss of the *Titanic*, the first international telegraphy conference was held in London in July 1912, and although its declarations were suspended for the First World War, they came back into force in 1919. More telegraphy conferences were held in 1920, 1927 and 1931. The first conference on the Safety of Life at Sea (SOLAS) was held in London in 1914. More conferences followed in the aftermath of the First World War, and each made important additions to the regulations. In 1912, the third international buoyage conference was held in Russia. By then the shape and colour of channel and middle ground buoyage had been agreed upon, along with fairway and landfall beacons. The design and location of buoyage to mark isolated dangers, moorings, anchors, wrecks and telegraph cables was also settled. To meet the new regulations for a ship's lifeboats that were introduced at the beginning of the century, collapsible life rafts were invented to save space but still meet the legal requirements. So, while this period is characterized by wars that made seafaring more dangerous than ever before, the sea itself was also being regulated for the first time; it was becoming safer.

The *Lusitania*

7 May 1915

The torpedoing of the liner Lusitania *off southern Ireland, which resulted in a huge loss of life, was a clear act of aggression widely condemned at the time. With hindsight, it represents a seminal moment in the history of conflict. The attack itself showed that the world was now engaged in a new mode of warfare, where chivalry had been supplanted by 'total war', involving non-combatants. Its aftermath was also a very modern affair, as propagandists and the press fomented popular outrage. One immediate result of the sinking was to turn American public opinion decisively against Germany, paving the way for US involvement in the First World War.*

It is not widely known that the great Cunard passenger liners, the *Mauretania* (launched 1906) and her sister ship the *Lusitania* (1907), were built with money lent by the British government at a very low rate of interest, and the vessels were designed for conversion to fighting ships at short notice. The *Lusitania* herself carried emplacements for 12 six-inch quick-firing guns. The Cunard Line then received a substantial annual subsidy to hold these vessels in readiness for war. In August 1914 that war arrived, and not long after the declaration, the *Lusitania* was briefly commandeered by the Admiralty, but soon returned to the Cunard Line when it was discovered how much coal she used. She reverted to her former role of transporting passengers, mail and freight across the Atlantic and retained her status as a merchant ship.

Even as a merchant vessel, however, she was vulnerable in those months. Although she was carefully searched by the special 'neutrality squad' before she left New York, which certified that she carried no armament, her previous associations with the Admiralty had ensured that her silhouette – the means by which a ship was identified from the periscope of a U-boat – had appeared in *Jane's Fighting Ships* for 1914, and both the *Lusitania* and the *Mauretania* were categorized as 'armed merchantmen' in the *British Naval Pocket Book* of the same year. The key threat to her security came in February 1915, however, when the German Admiralty declared a war zone around the entirety of the British Isles and stated that any enemy shipping discovered there would be sunk. Such was their declaration, but it remained illegal under international law to sink a merchantman without warning, and without endeavouring to save its crew.

This posed significant problems for the U-boat captains. To stop a merchant ship required the small, slow and unseaworthy U-boat to surface. This made U-boats very vulnerable. Their top speed was little more than 12 knots, whereas a fast merchantman or warship could achieve twice that, and could easily run down and ram a U-boat, a tactic which British merchant captains were explicitly instructed to use. (While serving as a troopship, the RMS *Olympic*, sister ship of the *Titanic*, rammed and sank the *U-103* in the English Channel on 12 May, 1918, the only sinking of a German submarine by a merchant vessel during the First World War.) Surfacing also left the U-boat vulnerable to the merchantman's guns or those of her naval escort, if she was fortunate enough to have one, and just one well-placed shot from a light gun, moreover, would be enough to sink a U-boat.

Unrestricted Submarine Warfare

At the beginning of the war, therefore, these legal restrictions on sea warfare led merchant captains to believe that they were free from the U-boat threat, and to a large extent they were. In the early stages of the war the U-boat targets were purely military and focused on the British North Sea fleet which blockaded the German coast, and on the British supply lines that fed the armies in northern France. However, after nearly eight months of war, the effect of the British blockade and the stalemate in France led the German navy to the new and desperate measure of unrestricted submarine warfare. If they did not grasp the initiative, it was clear that a slow economic death would stifle the German war effort. Their only option was to retaliate in kind and prevent Britain from growing fat with supplies and munitions. Britain could not be 'blockaded' in the traditional sense, however. The British bottled up Germany from Atlantic trade by blocking the English Channel and the entrance to the North Sea north of Scotland. The German navy had no such geographical advantage over her enemy and so British merchant ships themselves became the target. This was no passive blockade; this was an offensive war against British trade. This strategy was clearly incompatible with international law. In the eyes of the British and Americans this therefore made such a strategy unlikely; in fact it made it barely conceivable. In the eyes of the Germans, however, survival rendered international law redundant and so they began to train their submariners in the inhumane rigour of unrestricted warfare. U-boat decks were strengthened to carry heavy guns, which could be used against unarmed and unescorted merchant ships; bullets and shells were a cheaper way of destroying ships than torpedoes. Living space was reduced to increase storage for ammunition. Training was given to maximize the effectiveness of the U-boats. They were, for example, ordered to leave the vicinity of any ship slowly settling in the water or burning on the surface. U-boat crews took with them merchant captains who could advise on the type and nationality (and therefore suitability for destruction) of the vessel encountered. Warnings were not to be given. This was no half-hearted strategy. The German war machine was committed absolutely to the systematic destruction of all British non-combatants. Neutral ships and hospital ships were the only exceptions, but it was made quite clear by the German Admiralty that '*if, in spite of great care, mistakes should be made, the commanding officers will not be held responsible*'.

The British, at least, knew that the onslaught was coming, but they had no real idea how to meet it; remember that this was an entirely new way of waging war. Nevertheless, a great deal of energy was put into meeting this threat. Mines were laid and huge nets towed by drifters into the narrowest stretches of the English Channel. The mines were largely ineffective, however. Most sank or drifted away, and the nets were easily avoidable. In fact the German navy, now unencumbered by legal restriction, were hugely emboldened and treated these naïve attempts to stop them with the utmost contempt. They scouted for their targets, lurking near busy shipping lanes and close to shore where they were more sheltered from the elements.

The first U-boat sortie in February 1915 saw 11 British ships attacked and seven sunk. The hospital ship *St*

Andrew was also attacked, but she managed to escape. A French vessel was damaged and so too was a Norwegian tanker, the *Belridge*, which was loaded with American oil and destined for the Dutch government. The Americans watched with growing alarm. In March, 27 British ships were sunk, and for the first time one of them, the *Falaba*, bound for West Africa, was an unarmed passenger liner. Shortly afterwards, the *Arabic*, a White Star liner, and a French Line passenger ship, both came under attack, albeit unsuccessfully. In April the German embassy in Washington began to publish warnings in the press that British ships were liable to destruction, and that '*travellers sailing in the war zone on ships of Great Britain or her allies do so at their own risk*'. The German ambassador claimed it was a friendly warning. By the end of April, just one day before the *Lusitania* was scheduled to weigh anchor in New York, a total of 66 merchantmen had been sunk by German U-boats and mines.

In that month 18 vessels, 11 British and six neutral, were sunk almost exclusively by only two U-boats. There were some examples of counterattacks, armed merchantmen targeting U-boats or ramming them, but they were few. Look-outs had an almost impossible task in trying to spot a periscope, and the newly invented hydrophone, which allowed the user to 'hear' underwater, was not widespread, nor was its potential use against U-boats well understood. Effective depth charges, moreover, were not introduced until 1916. In the first few months of 1915, therefore, the crews of the merchantmen were blind and crippled. The best defence that the Allies could come up with at this time was to paint some of the merchant vessels in dazzle-camouflage to break up their silhouette against the horizon. This, it was hoped, would make them harder both to identify and to target.

Unfolding Tragedy

The American press teemed with news of British and neutral vessels being sunk, and these reports were printed alongside the German embassy's warning, but few paid them any heed. The *Lusitania* left New York at noon on 1 May with 1959 passengers, of whom 159 were American citizens. The voyage was largely uneventful, but the war around the British Isles during those few days certainly was not. In particular, one U-boat that had clearly escaped from the North Sea around the top of Scotland was leaving a trail of destruction as it progressed down the western coast of the British Isles and into the Western Approaches. On 28 April a collier was sunk off the Butt of Lewis, the next day another off the coast of Mayo. The day after that, two more were sunk off the southwest coast of Ireland and again the next day, on the very day that the *Lusitania* set sail from New York, three more were attacked off the Scilly Isles. Attacks continued on 3 May and 5 May, and two days later the *Lusitania* approached the Irish coast. Not all of these merchantmen had been sunk by the *U-20*, the submarine that was to sink the *Lusitania*, but it was clear that the waters around the Western Approaches were far from safe.

The movements of the *U-20* can be followed in great detail because the diary of her commander, Walter Schwieger (1885–1917), survives. Owing to a lack of fuel, he was unable to venture as deeply into the Western Approaches as he would have liked, but was patrolling the waters just to the south of the entrance to the Bristol Channel. With only three torpedoes left he was determined to use one of them and save two for the journey home. On the morning of 7 May his chance came. Just a dozen miles off the Old Head of Kinsale in southern Ireland, he spotted through his periscope a distinctive and large passenger liner with four funnels. There is no evidence that he recognized her for what she was, nor is there any evidence that he was ordered to lie in wait for that particular vessel, only to attack '*transport ships, merchant ships and warships*'.

When Schwieger first sighted the *Lusitania* he had no clear shot and the ships at that time were too far apart for Schwieger to force an interception. All he could do was position himself in the hope that she would turn to starboard along the coast of Ireland, and then continue slowly in a straight line towards him, which is exactly what she did. Schwieger's diary records what followed: '*Until 3 p.m. we ran at high speed in order to gain position directly ahead. Clean bow shot at a distance of 700 metres (G-torpedo, three metres depth adjustment); angle 90°, estimated speed twenty-two knots. Torpedo hits starboard side behind the bridge.*' There then followed a secondary, and enormous, explosion. Schwieger goes on to detail the effect of his attack in the clipped tones of a reporter observing an event he had just stumbled across, instead of one he had just caused.

> '*The superstructure right above the point of impact and the bridge are torn asunder, fire breaks out and smoke envelops the high bridge. The ship stops immediately and heels over to starboard very quickly, immersing simultaneously at the bow. It appears as if the ship were going to capsize very shortly. Great confusion ensues on board; the boats are made clear and some of them are lowered to the water. In doing so great confusion must have reigned; some boats, full to capacity, are lowered, rushed from above, touch the water with either stem or stern first and founder immediately...the ship blows off [steam]; on the bow the name* Lusitania *becomes visible in golden letters.*'

Schwieger immediately dived to 24 metres (79 ft) and headed for open water. When he returned just under an hour later, the *Lusitania* was nowhere to be seen and only a handful of lifeboats drifted in the mist. By then the liner had been sunk for almost half an hour. The German warnings had not been heeded and the *Lusitania* had sailed with her largest eastbound passenger list of the year. One thousand, one hundred and ninety-eight souls went down with her, including 270 women, 94 children and 128 American citizens – that is, just over two-thirds of all the Americans aboard. A German sympathizer wrote in the *New York Times* three days after the sinking that these people had, in effect, committed suicide. In reality, most believed that the speed of the *Lusitania* would fox any U-boat; this was no sluggish tramp ship but the liner that had once held the Blue Riband, the prize for the fastest crossing of the Atlantic. Others thought, quite wrongly, that she would be escorted by the Royal Navy as she approached dangerous waters. Most simply believed that the Germans were bluffing, and that they would never target a ship packed with civilians. The war to them was a distant horror, a matter of headlines and casualty lists. It was experienced vicariously through newspaper reports, not through the deliberate murder of their children.

'An Act of Piracy'

The world wept and the Germans squirmed. They claimed – falsely – that the *Lusitania* had been carrying Canadian troops. Another claim was that she had been carrying

British arms and munitions. This was true, but irrelevant; the type of cargo had no bearing on how she could be attacked. Finally, they claimed that the *Lusitania* had been warned not only by Germany's declaration of a war zone around the British Isles, but also by the notice in the American press and by the reports of merchantmen being sunk in the previous three months. The British and Americans had no grounds at all for complaint, argued the Germans.

Those who buried the remains of the bloated corpses in the mass graves (which can still be seen today) in the cemetery at Queenstown (Cobh) begged to differ. One sailor was recovered with the body of a little child strapped to his shoulders. Most of the millions worldwide who read the sensationalist reports in the press also rose in uproar. One Italian journalist wrote: '*We must go far back in time to find such a total suppression of humanity. Germany thinks to terrify the universe by filling it with as much blood as it is possible to spill.*' The Pope was roused to speak out and sent a personal cable to the Kaiser. The Germans knew that the line had been crossed and the Kaiser personally reprimanded Schwieger, although it is pertinent that his reprimand was deeply resented within the German submarine corps.

Schwieger was not the only commander to be reprimanded for the events of that day. The *Lusitania*'s captain, William Turner (1856–1933), had followed none of the three specific recommendations made to merchant captains on how to behave in U-boat-infested seas: they were to steam at top speed, stay in deep water, away from headlands and steer a zig-zag course. Schwieger intercepted the *Lusitania* sailing in a straight line under reduced speed, near the headland where the U-boat was seeking shelter from the Atlantic swell.

In the final analysis, therefore, Turner should not have made the *Lusitania* such an easy target. At a higher level, the British should have done more to protect her and some argue that they should not have loaded her with munitions to make her a tempting target. Once struck, the inexperienced crew failed to react swiftly and efficiently, although this is not surprising as many of her experienced crew had been conscripted. There had, moreover, only been one lifeboat drill on the voyage. The statistics reveal that many of the crew may well have sought to save themselves rather than offer any assistance to the passengers: 41 percent of the crew were saved, but only 37 percent of the passengers. Unlike the *Titanic*, fatalities were distributed evenly among the three classes of passenger aboard, but the survival rate of men, women and children was unequal. More men survived than women, but only a little over 27 percent of the children aboard lived to tell the tale.

These were the technical arguments that raged, but underlying it all was a case of mass murder that could not be ignored. Former US president Theodore Roosevelt railed against the attack as *'an act of piracy'*. Although the sinking of the *Lusitania* did not directly lead America into the war (President Woodrow Wilson merely sent four diplomatic notes of protest), it did without any doubt form a significant part of that web of events that gradually eased America towards an alliance with Britain. Most significantly, it led to the complete collapse of the German propaganda effort in America and a temporary suspension of the unrestricted U-boat campaign.

Almost two years after the sinking of the *Lusitania*, and just when the renewed U-boat war was starting to show signs of victory, with German strategists predicting an economic collapse in Britain within six months, America declared war on Germany in April 1917. American troops flooded into France. Almost two million men were landed in France and kept there with unbroken lines of trans-Atlantic supply. The Americans also took over many of the Royal Navy's Atlantic duties in mine laying, convoy escort and U-boat patrol, which freed the navy to concentrate on the threat from the German surface fleet in the North Sea. The Germans never recovered from this imbalance in material resources and the war soon ended.

A Sorry Sight

Since the *Lusitania* sank in a relatively well-known position, in relatively shallow water, she has attracted a great deal of attention from treasure hunters and wreck divers. By the 1980s three of the four bronze propellers and two bow anchors had been taken. In 1990 one of the propellers, which had been removed with plastic explosives, was then melted down and turned into 'commemorative' golf clubs, selling for $9000 a set. There are few more shocking examples of blatant commercial exploitation of the world's maritime heritage.

She was then explored in 1993 by Robert Ballard (b.1942), the man who found the *Titanic* in 1985, and since then a number of other expeditions have studied her remains more closely. Thankfully, she is now protected by an Irish Underwater Heritage Order, which forbids unauthorized diving. We know that when she sank she collapsed onto her starboard side, obscuring the great rent caused by the torpedo and the secondary explosion. Unlike the *Titanic*, her hull's infrastructure has almost entirely collapsed and she is pocked by depth charges from naval exercises and unkindly decorated by snagged fishing nets. Even in appearance she is a sad wreck, but all stories of ships that are deliberately sunk by man with loss of life are deeply distressing. Many believe the tale of the *Lusitania* to be the saddest of them all.

Scapa Flow

21 June 1919

As the world awaited the outcome of the Versailles peace negotiations ending the First World War, an extraordinary event unique in naval history was played out in the Orkneys, north of the British mainland. After eight months' squalid internment, the admiral commanding the German High Seas Fleet ordered his men to scuttle their 74 vessels to prevent their use by the victorious Allies. This unprecedented act of self-destruction restored the pride of the Imperial German Navy, but most of the sunken ships were salvaged and their steel reused in new British warships, as the next global conflict against a resurgent Germany loomed on the horizon.

Two weeks after the end of the First World War, the British Grand Fleet and the German High Seas Fleet, two of the biggest fleets of warships ever to have existed, met for the second time. Their first encounter had been at the Battle of Jutland at the end of May 1916, and it had been a fierce clash. Both sides claimed victory, and although more British ships (14) were sunk than German (11), many more German ships than British were sufficiently damaged to be forced into dock. The impact of the battle on each navy's subsequent strategy is most telling of its outcome: while the British strained at the leash to meet the Germans in pitched battle again, the Germans knew that they could never again risk such an open engagement with the Royal Navy. The Imperial Navy focused almost all of its attention on a submarine war against British trade. The surest way to cripple Britain, they realized, was to starve her into submission, not to confront her navy in open, surface warfare.

Since Jutland, therefore, the German fleet had influenced the war through its presence rather than through its activities. Only two hesitant sorties were made, both to little effect, but the High Seas Fleet's very existence, glowering over the eastern horizon of the North Sea, protected the German coastline and the Baltic from Allied interference and helped secure the U-boat bases that the Germans believed were the key to victory. Now the two fleets met once again in the aftermath of that war; a war that Germany had irrevocably lost. The German nation was ravaged by revolution, her population was starving, thousands of her soldiers were dead, and the emperor had been forced into exile. This time, the two fleets did not meet to give battle, but to join and return to the British naval base at Rosyth for inspection, before continuing their journey to the anchorage at Scapa Flow in the Orkneys. There, the German fleet was to be interned while the great powers met at Versailles to negotiate the details of the armistice.

The Final Voyage

The meeting of the fleets had been carefully stage-managed by David Beatty (1871–1936), commander-in-chief of the Grand Fleet, for maximum effect. If the German fleet was not to be defeated at sea in decisive battle, as the navy had hoped, then their internment was to be made as symbolic as possible; their internment was to be that defeat. Accordingly, a few days before the two fleets met, the German Rear-Admiral Hugo Meurer (1869–1960) was invited to Scapa Flow aboard a cruiser to 'negotiate' with the British, although in reality the Germans were in no position to manoeuvre. Meurer was transferred to Beatty's flagship, the colossal super-dreadnought *Queen Elizabeth*, one of only four *Queen Elizabeth*-class battleships, the most powerful type of her day. Meurer was met there by a flood-lit double file of hand-picked extra-tall Royal Marines in full-dress uniform with fixed bayonets. There he met Beatty, who treated the occasion as a full surrender, rather than the internment, which, legally speaking, it was.

*

Six days later, the two fleets met on 21 November, 1918. The German fleet of 70 warships, all steaming in line-ahead formation, was over 19 miles (30 km) long. The British, over 100 strong, flanked them in two parallel lines, with all guns loaded and trained, and they shepherded their prize to the Firth of Forth. Under British inspection the Germans landed their ammunition and removed the breech blocks from the great guns, rendering them impotent at a single stroke. From there the High Seas Fleet was escorted to Scapa Flow, a prison of its own. Once they had arrived, David Beatty made the following signal: '*The German flag will be hauled down at sunset today, and will not be hoisted again without permission.*' The Germans were interned, but in the eyes of the world this was unequivocal surrender. The distinction was significant both for the British and the Germans. For the Royal Navy, to treat the handover as surrender was to claim a victory; for the Germans, to surrender was the ultimate dishonour.

A Desolate Anchorage

Scapa Flow is one of the finest natural harbours in Europe. Over 311 square kilometres (120 sq mi) in area, it has direct access to both the North Sea and the North Atlantic, and is almost completely surrounded by small islands that protect it from some of the worst weather in the northern hemisphere. Its depth as well as its magnitude allows vessels of any size to seek shelter there. It is also a very hostile environment. The surrounding hills are low and windswept. There are very few trees anywhere on the nearby islands. The soil is poor, the weather frequently and bitterly cold, and plagued by fog and rain. So far north, the winters are characterized by very short hours of daylight, and in the summer an equally small amount of darkness. It was so remote that one British newspaper reporting on the internment of the Germans got its position wrong. British sailors unanimously loathed it. One of the popular songs sung by the sailors of Scapa was called the 'Hymn of Hate'. Such unity in their dislike for the place created a strong feeling of comradeship which made it bearable.

The British, at least, did what they could to improve conditions. The island of Flotta was designated as the fleet sports ground, where regular games of football were held. Three rough golf courses were laid out for use by the officers. Ash and gravel tennis courts were made. There were regular sailing and pulling races around the fleet. The annual Grand Fleet Boxing Championship was witnessed by as many as 10,000 men. Some ships grew vegetables ashore while one ship's company even managed to raise chickens and pigs, although the unpredictable nature of a warship's routine made the survival of these poor animals rather uncertain.

The Germans, on the other hand, were not allowed any luxuries. Their radio receivers were removed by the

British, and their only access to newspapers was through those distributed by the Admiralty. The German sailors who were fortunate enough to be able to read English passed the news on to their shipmates, but it was always at least four days old. They were not allowed ashore at any time, and all of their food had to come from Germany, which was suffering badly from the effects of four years' effective maritime blockade. German economic infrastructure had almost totally collapsed and the entire population of Germany, and therefore her fleet at Scapa Flow, lived almost exclusively on turnips. The sailors augmented their frugal diet with the fish that they could catch from their boats, usually tiny sprats which swam in swarms around the hulls, presumably feeding on the sewage that was pumped out.

To make conditions easier the crews were reduced to a bare minimum, and the excess men sent back to Germany. Those that remained did little more than the basic maintenance required to keep their ships afloat. Without regular and structured care the ships rusted as quickly as the men became ill. The men's teeth started to fall out and the hatches on the ships became seized shut. The ships were overrun with cockroaches and rats. The Imperial German Navy, pride of the Kaiser and tool of his imperial ambition, rotted at Scapa from the inside as quickly and completely as his vision of a vast German empire.

A Spectacular Act of Sabotage

The terms of the armistice, meanwhile, had been decided. The Germans were allowed to retain a fleet, but one that was only a fraction of its present size. None of those held at Scapa would be returned but they would be allowed to build a new fleet of 6 battleships, each not more than 10,000 tons; six cruisers, each less than 6000 tons; and 12 torpedo boats. If the Germans did not accept these terms, hostilities would recommence. They had four days to consider their response. The German admiral in charge of the High Seas Fleet, Ludwig von Reuter (1869–1943), believed that the terms would never be accepted by Germany, and although he had more time to wait for an official response than he believed (the deadline for acceptance was extended for another two days), he took the only action he believed was available to him, and scuttled every last one of the 74 German ships in Scapa Flow.

When the order was given, every torpedo tube, valve or pump that led from the hull directly to the sea was opened aboard every ship in the German fleet. Every watertight bulkhead was opened, the hinges jammed and locks broken. The valves that allowed the ships' various compartments to be pumped free of water were disconnected. Everything had been prepared for almost a month, and they sank like stones. Aboard every ship the German ensign flew one last time in defiance of her captors.

This tragic end of these great ships, as they plunged beneath the waves, hissing, groaning and collapsing as the weight of the water filled their hulls and drove them towards the sea bed, was only witnessed by a handful of people. The entire Grand Fleet had sailed that morning for gunnery exercises in the Pentland Firth. Left in Scapa Flow was a tiny squadron of naval ships, the guard force of the German High Seas Fleet, which included the Admiralty tender *Flying Kestrel*. And aboard the *Kestrel*, in the very midst of the German fleet, was a group of around 400 schoolchildren from Stromness who were being taken on a tour of the German fleet. Sick with excitement, they were finally brought alongside the German ships, whose decks were lined with sailors '*who did not seem too pleased to see us*', as one of those children, James Taylor, later recalled. This same eyewitness then carries on the tale:

> *'Suddenly, without any warning and almost simultaneously, these huge vessels began to list over to port or starboard; some heeled over and plunged headlong, their sterns lifted high out of the water. Out of the vents rushed steam and oil and air with a dreadful roaring hiss. And as we watched, awestruck and silent, the sea became littered for miles round with boats and hammocks, lifebelts and chests ... And among it all hundreds of men*

struggling for their lives. As we drew away from this nightmare scene we watched the last great battleship slide down with keel upturned like some monstrous whale.'

We know from contemporary diaries that to witness one single ship of this size disappear beneath the waves was a profoundly shocking experience. To witness 74 such ships in their death-throes was almost impossible to comprehend. It is curiously fitting that this one final tragedy of this terrible war, a war that had killed more men faster than any other, was not witnessed by hardened veterans of conflict, by men who had already witnessed fully-manned warships explode before their eyes, but by children. The war had savagely beaten the innocence from the world, and its final act was to scar the memories of those children from Stromness.

Only seven sailors actually died in the process of scuttling, so well planned was the operation. Indeed, most of those who died were killed by British bullets as the guard force desperately tried to limit the damage. One battleship, three cruisers and eleven torpedo boats were towed to shallow water where they could easily be refloated. Two thousand men needed immediate shelter, food and medical care before they could be transported to a secure environment as prisoners of war. The surrendering Germans were treated with 'minimum courtesy'; there are reports of many being assaulted and robbed. They were temporarily housed aboard three British battleships. When they finally returned home in January 1920 they received a hero's welcome as men who had salvaged the pride of the German navy. In July 1919 Admiral Reinhard Scheer (1863–1928), commander of the German fleet at Jutland, wrote in a letter to *The Times*: '*I rejoice over the sinking of the German Fleet in Scapa Flow … the stain of surrender has been wiped out from the escutcheon of the German Fleet.*' While the economically and militarily crippled Germany now dealt only in the abstract concept of pride, British concerns were tangible; they were of steel, steam, guns and shellfire. In their eyes, the pride of Imperial Germany was its fleet, and it had been lost. To them, the scuttling was shameful; an inglorious end to a magnificent fleet. A shining tool of empire had become a rusting source of scrap metal.

The impact of the scuttling was significant in other ways too. Before it was destroyed, the German fleet was a significant pawn in the power struggle that evolved in the aftermath of the war. British naval supremacy was already absolute, and if it was augmented by the German fleet then it would be untouchable for many years to come. If, on the other hand, the German fleet was divided between America, France and England, then the dominance of the British would immediately be reduced. The destruction of the fleet now left the situation much as it was before. The French, in particular, were furious, and even accused the British of complicity in the sinking. The Americans on the other hand, were not overly concerned. They would certainly have liked some of the ships, but the next-best solution was for them to be denied to the British. That way, the struggle for naval dominance would be reduced to a competition between industrial shipbuilding infrastructure, and in the long run the Americans knew that they would win that race with their greater natural resources.

Salvaging the Fleet

After the war, the wrecks at Scapa became a peculiar resource of their own, and private companies began to salvage them for scrap metal. By the 1920s, the British government paid huge sums for scrap metal to help supply the shipbuilding industry as the world re-armed. The salvage operations at Scapa were some of the most daring ever attempted. Numerous techniques were used, but the most spectacular, pioneered by the London scrap-metal dealer Ernest Cox, involved sealing all of the holes in the ship before pumping the hull full of air. This lifted the ship off the bottom ever so slowly, but as it rose, the air expanded under the decreasing ambient pressure, and the ship ultimately launched herself through the surface. It is ironic that the only salvage platform large enough to cope with some of these warships had to be imported from Germany: a German salvage platform was used to raise German warships, which were being cut up to build

British warships – so that they could, once again, wage war with the Germans.

Most, but not all, of the German wrecks were subsequently salvaged, but the remains of the German ships are not the only wrecks left today at Scapa. To protect the anchorage, a number of block ships were sunk at the entrance, and because of the increased circulation of water that their location affords, they are much more visible underwater than are the battleships sunk in the deep and still waters of the sound. The wrecks of two British battleships also lie in Scapa: on 9 July, 1917, when a stokehold fire on board HMS *Vanguard* ignited a store of cordite for her guns, she exploded with such force that an entire 400-ton 12-inch gun turret landed on Flotta, a mile away; 804 men died. In the early months of the Second World War, HMS *Royal Oak* was torpedoed by a German U-boat in the very heart of Scapa Flow; 833 men died.

With its calm and relatively clear water, Scapa Flow is one of the finest wreck sites in the world. Nowhere else on Earth are examples of the first breed of dreadnoughts so accessible, and nowhere else is there so much wreckage on the seabed: 3 German battleships, 4 German light cruisers, 2 German torpedo boats, 2 submarines, a Second World War German 'escort boat' (*Geleitboot*), 2 British battleships, 16 other British wrecks, 27 large sections of salvors' equipment, 32 block ships and 54 more unidentified pieces of wreckage.

Final Resting Place

The men who lost their lives in the various actions and accidents at Scapa Flow can all be remembered in the beautiful naval cemetery at Lyness, where the few German graves are tended as carefully as the many British graves. More than 2000 men died at Scapa Flow in the course of the two world wars, but their bones lie beneath the flowering heather, while the skeletons of their ships haunt the mirror-calm waters of the bay. There is very little else there that provides clues to the hidden history of Scapa. Apparently unspoiled, untarnished by the hand of man, this beautiful landscape conceals human history of the worst kind. So remote from modern society and yet so pivotal in 20th-century history, few places in the world can match Scapa Flow for its atmosphere and troubling secrets.

USS *Arizona*

7 December 1941

The Arizona *was sunk with an appalling loss of life during the Japanese attack on Pearl Harbor on 7 December, 1941, described by US president Franklin D. Roosevelt as* 'a date which will live in infamy'. *Her hull is now a war grave and national monument, and yet the fleet of which she was part was carefully salvaged to become the chief instrument of America's final victory over Japan.*

The USS *Arizona* does not conform to one's impression of what an internationally recognized symbol of maritime endeavour should look like: this is no USS *Constitution*, whose beautiful rigging laces the Boston skyline, nor is it a ship like HMS *Belfast*, whose stern and hostile silhouette breaks up the London horizon, her very presence an uncomfortable splinter in the soft underbelly of the British capital. The *Arizona* is none of these; she is not even afloat. Indeed, the casual observer would see no warship at all where she lies; all that is visible of her from the surface is a rusting circle of steel – the bottom of a gun turret – that pokes out of the water. And yet the *Arizona* is visited by more than 1.5 million people each year. All go to learn about America's maritime past, but for some this is no idle voyage of curiosity, but a pilgrimage of great significance. For the *Arizona* is far more than just a symbol of US maritime heritage: it is the tombstone and war grave of over 1000 men; it is a shrine to both defeat and victory; it is the spark that once burned so fiercely that it led to the creation of the largest and most powerful navy the world has ever seen. She is, without a doubt, one of the most significant wrecks in the world.

Victim of an Infamous Attack

The *Arizona* was sunk at her berth in Pearl Harbor, on the Hawaiian island of Oahu, by the successive waves of Japanese bombers that attacked the US naval base there on the morning of 7 December, 1941. Struck numerous times, it was one single 800-kilogram (1760 lb) bomb that finally destroyed her. It pierced her teak decks near No. 2 turret towards the bows, and then burst its way through the decks until it landed in the forward powder magazine, where nearly a million pounds of powder was stored. The resulting explosion peeled back the steel of the bows and vapourized everything below No. 1 turret, which fell vertically almost 9 metres (28 ft) and landed in the correct orientation, but now resting on the stricken ship's keel. Her ruptured fuel tanks bled oil, which immediately ignited. Now a funeral pyre and emergency beacon, she burned for three days as the hull slowly settled in the mud. One thousand, one hundred and seventy-seven of her crew died on that day, 1100 of them entombed inside her as she sank.

It was almost immediately clear in the aftermath of the attack that the *Arizona* was too badly damaged to be salvaged, and by the time the fires on Oahu had been put out, the wounded attended to, trapped survivors freed, as much firepower as possible resurrected and the men prepared for another attack, many days had passed. There had been no time to recover the dead from the *Arizona*, nor indeed from many of the other ships with an equally sinister cargo. When that grim task was eventually undertaken, the bodies were bloated, much of their clothing had rotted away, and their extremities been reduced to sinew and bone. In the oppressive tropical heat of Hawaii those corpses that were raised to the surface emitted such a stench that several medical orderlies responsible for bagging up the dead vomited into their gas masks. The task of recovering the bodies from the *Arizona* was quickly abandoned; her dead would remain where they lay. Meanwhile, the ship itself was stripped of anything useful (guns and ammunition), confidential (the ship's safe) or personal.

Prelude to Tragedy

The events that led to the destruction of the *Arizona* are well known. By September 1940, America and Japan were set on a course for war. Japan had recently signed the Tripartite Pact with Italy and Germany, aligning herself with the fascist Axis Powers that were preparing to ravage Europe. Tokyo, meanwhile, was determined to realize its dream of a 'Greater East Asia Co-Prosperity Sphere' – in other words, a Japanese empire dominating the whole of Southeast Asia. British, Dutch and French colonies in the region were all weakened as their host nations turned their backs on the Pacific to face the German threat in Europe. America reacted to Japanese aggression by blocking Japanese advances where possible, issuing a strict embargo against all industrial exports to Japan, and seizing Japanese assets in the United States. Diplomacy between the two countries continued in an atmosphere of mutual mistrust.

American mistrust, however, did not extend to the belief that the Japanese would attack them at Pearl Harbor, the base of the American Pacific Fleet. Oahu was a citadel; the most heavily garrisoned island in the world, bristling with hundreds of aircraft, powerful coastal batteries and the mighty guns of those ships of the Pacific Fleet that were not out on exercise. It was also a tropical paradise, filled with tourists even in those troubled years.

Part of an isolated island chain in the depths of the Pacific, it was impossible to escape the impression on Oahu that one was cut off from the rest of the world. Although relations with Japan had been noticeably deteriorating, and the ships, men and aircraft trained to an ever-increasing degree of readiness for conflict, there had been no declaration of war. Many believed that war would come, but not now, not yet, and when it did come, everyone was sure that it would be fought in countries far away. On the morning of 7 December, 1941, on heavily protected American soil, everyone on Oahu thought that they were safe.

The Japanese perspective was very different. Their fleet of aircraft carriers, the most powerful in the world, was manned with pilots brimming with confidence, their skills honed in the recent war against China. The Imperial Japanese Navy was also led by Isoroku Yamamoto (1884–1943), a man steeped in his country's brief, but very impressive, naval history. In 1894 a Japanese naval force initiated action in their war against China by sinking a Chinese warship and a troopship before war had been declared. Ten years later the Japanese navy launched an attack on the Russian Far Eastern naval base of Port Arthur, again before war had been declared. In both instances, the Japanese naval attacks had handed them the initiative in the war, and their enemy had never been able to recover. By the end of those conflicts, both the Chinese and the Russian fleets had been annihilated.

Yamamoto's plan was to combine this tactic of a pre-emptive strike with the new power afforded by aircraft launched at sea. The success of such an attack had been demonstrated by the Royal Navy's attack on the Italian naval base at Taranto in November 1940. In that attack, against a hostile, fully prepared and defended port, slow biplane Fairey Swordfish torpedo bombers launched from a single carrier had sunk or badly damaged three Italian battleships and one light cruiser. Yamamoto envisaged a force of six carriers, which, working together could launch more than 400 of the most advanced warplanes in the world. They would be assisted by five midget submarines, all armed with state-of-the-art torpedoes. Both would face an unsuspecting enemy. The carrier force itself would be protected by three submarines, a light cruiser, nine destroyers, two battleships and two heavy cruisers. It was the largest assemblage of sea power ever to have gathered in the Pacific. They would be almost impossible to stop. While this fleet cruised unnoticed towards American waters, aboard the *Arizona* in Pearl Harbor, her captain, Franklin Van Valkenburgh (1888–1941) wrote to his wife. '*By this time next week*', he said, '*we will be on our way home for Christmas*'. A day later he and over a thousand of his crew were dead.

Significance of the *Arizona*

Most obviously, the wreck of the *Arizona* recalls that attack, but it also has a far wider resonance. Firstly, it is significant that she is a battleship, and not an aircraft carrier. By 1940, maritime strategists had appreciated the potential of aircraft launched from ships for more than 20 years, and in that time, ships, aircraft, tactics and strategy had been developed to make seaborne air warfare a reality. The attacks on Taranto and Pearl Harbor were two of the most important events in the war thus far, both carried out by aircraft launched from ships. It was now impossible to gain control of the sea without exercising air power; and the years that followed would see sea-battles fought between fleets hundreds of miles apart. There would be few confrontations between battleships, and none on the scale of the Battle of Jutland in 1916. It was indeed fortunate, therefore, that none of the American Pacific Fleet's aircraft carriers were destroyed in the Pearl Harbor attack – the three carriers, the *Yorktown*, *Lexington* and *Saratoga*, were all out at sea at the time. Even after the attack, therefore, the Americans had a firm base from which to grow; they could control the sky around their remaining warships and they could project naval power anywhere on the globe. The ultimate success of the Allied war effort in the Pacific rested heavily on the role played by the American carriers.

It is also significant that part of the *Arizona*'s superstructure now breaks the surface. She lies on the seabed at the berth where she was secured on the morning of the attack – berth F-7 – one of the most central of the berths on what became known as 'Battleship Row', the line of moorings that serviced the pride of the American fleet. And yet she lies in only 11.5 metres (38 ft) of water. Even here, where the big ships tied up, Pearl Harbor is not deep; parts of Sydney Harbor, by contrast, are nearly 46 metres (150 ft) deep. This made the salvage efforts of the resolute navy far easier in the days following the attack. Although the work was back-breaking and highly dangerous, claiming a number of lives itself, the salvage was at least possible. In fact every single ship that was sunk, with the exception of the *Arizona* and the *Utah*, was subsequently raised. And of those that were raised, all but one re-entered service, having been repaired at the extensive and undamaged ship-repair facilities at the harbour. Those facilities, along with the island's stockpile of 4,500,000 barrels of fuel oil, were the targets of an intended third wave of attack, but that attack was never carried out by the conservative and cautious commander of the Japanese fleet, Chuichi Nagumo (1887–1944).

Much of the significance of the *Arizona*, therefore, does not lie in her presence *per se* but in the absence of other similar warships. America was shocked, but her navy was not crippled, and the infrastructure that supplied, provisioned, built and repaired it was untouched.

Preserving a National Historical Landmark

The need to preserve the wreck of the *Arizona* has posed conservators serious problems. Her hull is around 183 metres (600 ft) long and 30 metres (100 ft) wide and she is almost entirely submerged in a rich chemical and biological soup that is steadily eroding her. Marine biologists and archaeologists are working in tandem to understand the nature of sea life on and around the wreck and the impact that it is having on the hull. Much of this has been pioneering work; very little on the preservation of sunken steel was known when the *Arizona* became an historic site in the American National Parks Service in 1962, but the need to preserve her has accelerated the science of ship preservation. She stands out from many other preservation projects because she is already in the most important and most appropriate context; that is, on the seabed. There, the power of the wreck is manifest. To see the sunken wreck of a battleship so close, and from dry land, shakes our accepted norms; indeed it is a profoundly disorienting experience that echoes, however faintly, the shock of those who witnessed the attack. In the middle of a defended harbour, on an island in the middle of the Pacific, the sunken *Arizona* defies preconceptions of historic ships.

One cannot help but look to the sky and horizon and feel suddenly vulnerable and uncomfortable. Even if one is not Hawaiian or an American citizen, it is an extraordinarily intrusive feeling, and it awakens a potent and natural desire to protect one's family, home and land. That desire for protection is one of the basest urges for conflict, and it is why visiting the wreck of the *Arizona* is such a powerful experience: it does not just teach us about the past, but forms an emotional bridge between us and those who fought for what they held dear in the Second World War.

The *Wilhelm Gustloff*

30 January 1945

The liner Wilhelm Gustloff *was built in 1937 for the German state-run leisure organization* Kraft Durch Freude *('Strength Through Joy'), which organized holidays and other pursuits for the Aryan 'master race'. Strongly associated from the outset with the Nazi regime, the brief life of this ship was to mimic almost exactly Hitler's years in power. As the Third Reich neared its bloody end, the* Gustloff *was torpedoed and sunk as she evacuated refugees from Danzig. It was the greatest loss of life ever on a single vessel. Precisely three months to the day after her loss, Hitler shot himself.*

The wreck of the *Wilhelm Gustloff* is, by some margin, the worst disaster in maritime history. No one knows exactly how many men, women and children went down with her, but the general consensus is around 9000. That is almost six times the casualty list of the *Titanic*. Some may remember the *Lusitania*, the *Athenia*, the *Andrea Doria* or the *Empress of Ireland* as powerful examples of mass death through shipwreck, but more died on the *Gustloff* than all of those wrecks put together.

From those figures alone the *Wilhelm Gustloff* deserves her place in any history of shipwrecks, but the context in which the wreck happened also makes it one of the most symbolic, rivalling the loss of the *Mary Rose* or the *Vasa*. More modern parallels of highly emblematic shipwrecks are the German battleships *Bismarck* and *Tirpitz*, the British *Hood* and the Japanese *Yamato*. Each in their time was the pride of their fleet, and each was destroyed by enemy action. The *Gustloff*, however, was no warship. Originally a passenger liner, at the outbreak of the Second World War she became a hospital ship, before being used as a troop transport, then as a floating barracks, and lastly as a refugee ship. This does not read like the *curriculum vitae* of an important ship, but to understand her full significance one must first consider the identity of the man she was named after. And the best way to do that is through the eyes of David Frankfurter (1909–82), Gustloff's assassin.

Frankfurter was a Jewish medical student outraged by increasing Nazi anti-Semitism in the 1930s. He believed that the Jewish people should not tolerate such abuse, that they needed to stand and fight, and he was prepared to lead them into battle. On 4 January, 1936 – *Ki Tov*, a Jewish day for good luck – Frankfurter murdered Wilhelm

Gustloff, the leader of the Swiss Nazi Party, in his own sitting room. His body, riddled with bullets, collapsed beneath a signed portrait of Adolf Hitler on a wall emblazoned with Nazi paraphernalia. Gustloff was the first Nazi leader to be assassinated by a Jew.

Flagship of Nazism

February 1936 was a time of great activity in Germany. Hitler's rise to power was complete, he was planning to march German troops into the demilitarized Rhineland and Berlin was preparing for the summer Olympic Games. The Nazi party preached its doctrine to the many thousands it had enthralled, and the latest propaganda coup was the construction of a magnificent passenger ship. This was not a ship that was built only for the rich, or one that emphasized wealth, class or professional distinction. The *Gustloff* incorporated large, unencumbered decks where there was room for everyone to sunbathe and for the passengers to interact with the crew. Every passenger cabin had a view, and was constructed to a uniform size; there was no cheap 'steerage' class confined to the bowels of the ship, as there had been on passenger liners of old such as *Titanic*. Indeed, there was no price distinction at all between tickets, which were roughly one-quarter of the price of voyages aboard similar European ships.

Moreover, she was not built to transport people from A

to B; this was no glorified ferry, but a true cruise ship as we understand them now. The ship was a holiday destination in itself. She was not the world's first cruise liner – the British shipping line P&O launched the first ship designed purely for pleasure cruising in Australia in 1932 – but she was the first such vessel in Europe, and her politically influenced design and shipboard routines made her even more remarkable. Each day was strictly organized into *Speisekarten* ('menus'), with associated music, games, sport and dance, while excursions to foreign ports were strictly controlled with coupons and other paperwork. The ship and her operation were carefully designed to ensure widespread acceptance of National Socialism and to raise the perceived standard of living of the German worker as German industry stepped into overdrive to arm for the coming war.

Now, however, the Nazis seized upon Gustloff's assassination as valuable fuel for their anti-Semitic policies. Before his death few people outside Switzerland had heard of him but he was suddenly raised aloft as a martyr to Nazism, an honourable man slain by Jews to spite the proud and righteous beliefs of the NSDAP. He was accorded a garish and heavily symbolic state funeral in which his coffin travelled through many towns, and was seen by many thousands of people, before arriving at the village where he was born. Hitler did not stop there. The magnificent ship on the stocks at Hamburg was personally renamed *Wilhelm Gustloff* by Hitler in yet another lavish ceremony. Her launch in May 1937 was accompanied by bands playing the *Horst Wessel* song (the battle song of the Nazi brownshirts, the *Sturmabteilung* [SA]), the national anthem *Deutschland über Alles*, and by Nazi flags and salutes. The launch of no other German ship had ever generated such public passion. This was not a warship that the German public would never see and could never board, that would carry their flag to distant countries and fight over the horizon; the *Wilhelm Gustloff* was a promise made good, she was physical proof of state investment in public leisure and health, and she was wildly celebrated.

Changing Roles

The first 17 months of her life went much as planned, as she provided over 65,000 Germans with an experience of a lifetime, cruising around the Norwegian fjords in the summer, and heading for Portugal, and then on to Italy in the mild Mediterranean winter, where her passengers could be assured of a friendly reception from Mussolini's fascist government. When war broke out in 1939, the German navy, the *Kriegsmarine*, was quick to requisition such an enormous vessel, and her first role was as a hospital ship.

She served in this role until the winter of 1940, when the German U-boat war was at its height. Following on from the success of submarine warfare against British trade in the First World War, the Germans poured resources into their U-boat fleet from 1939, placing great emphasis on the thorough training of new recruits. In November 1940 the *Gustloff* became the barracks of the Second Submarine Training Division at Gotenhafen (modern day Gdynia in Poland on the Bay of Gdansk). At this early stage in the war these new recruits would swell the ranks of the German U-boat flotilla, but within three years their role was simply to replace those who had died. Similarly, much of the German war machine began to crumble as Allied successes mounted. By October 1944 the tide had turned in northeastern Europe and the Red Army gained momentum in its brutal counter-offensive of revenge against German atrocities in the Russian 'motherland'. On 22 October, 1944 the Soviets overran the German village of Nemmersdorf. The Nazi propaganda machine immediately claimed, on very shaky foundations, that the Russians had massacred many of the inhabitants with spades and the butts of guns to save bullets. There were reports that some women had been stripped naked and crucified on the doors of their houses, and the skulls of their babies crushed.

The gruesome accounts immediately boosted German recruitment to fight the Russians, but they also created a ripple of panic throughout the eastern borders of the Reich. Thousands fled their homes in search of greater security in the west, and the Gulf of Danzig and the Vistula Lagoon (*Frisches Haff*) became a magnet for these refugees. Ships there would take refugees in Danzig, and other towns all along the coast, to safety. Known as 'Operation Hannibal', this was the largest wartime evacuation in history. Over 300,000 soldiers had been evacuated from Dunkirk in May 1940, but in Operation Hannibal over two and a half million soldiers and refugees were evacuated by more than 1000 merchant vessels. In April 1945 nearly 265,000 people were evacuated from Danzig alone. The *Wilhelm Gustloff*, lying at Gotenhafen only a few miles from Danzig, was immediately requisitioned. Nazi officials, prisoners of war, German soldiers, sailors and countless refugees crammed aboard. She was designed to carry 1465 passengers, but by filling the large public spaces on the open decks, as many as 10,500 people made it aboard. The temperature was −18°C. The sorely pressed *Luftwaffe* was unable to give the ship any air cover. Only two ageing torpedo boats provided protection from submarine attack, and one of those had to return to Gotenhafen almost immediately with engine trouble.

If there was any comfort to be taken from their plight, it was the inefficiency of the Russian navy. It is generally accepted that in the first few months of 1945, the German navy managed to evacuate troops and refugees from locations in operations that should have been wholly disastrous. The Russians made very little use of their surface vessels and their submarine force was small, out of practice and unwilling to take risks. Russian naval leadership was unimaginative, slow-moving and unaggressive. The captain of the Russian submarine *S-13* was not of this type, however; indeed he had a great deal to prove to his superiors.

A Fatal Encounter

Alexander Marinesko (1913–63) had been ordered to patrol the Baltic around the Gulf of Danzig from 2 January onwards, but his hard-living ways disrupted official plans. He was found drunk in a brothel the day the submarine was scheduled to leave. It was also known that he had spent New Year's night with a Swedish woman when relationships with non-Soviets were illegal. This was not the first time that Marinesko had come to the attention of the authorities, and his career had already been scarred by alcoholism. In February 1942 he had been expelled from the Communist Party. This was not behaviour that matched the Stalinist ideal, and Marinesko was arrested by the NKVD (the Soviet secret police, forerunner of the KGB) but released to demonstrate his loyalty.

He was told to patrol the coast near Memel, 155 miles (250 km) north of Danzig, but Marinesko chose instead to head south, to waters patrolled by German ships, where the risks and subsequent rewards were far higher. The

Gustloff, meanwhile, had turned her lights on to prevent collision with a fleet of minesweepers known to be in the vicinity. She was also painted naval grey. In her most recent life as a barracks ship she had been a legitimate military target and her priority had been camouflage. The startling white paint from her years as a hospital ship had been replaced and there was now nothing about her appearance to tell Marinesko, as he peered at her through his periscope, that almost two-thirds of her complement were women, children, injured soldiers and POWs.

Three torpedoes sped towards the *Gustloff*. The first struck her in the bows where the crew quartered. As the forward compartments flooded, the entire bow section was sealed off from the rest of the ship; everyone there died. The second torpedo hit amidships, at exactly the height of the ship's swimming pool, which, now drained, was being used as a dormitory for the German women's naval auxiliary service. Three of the 373 women escaped the deadly shower of tiles and glass that exploded as the torpedo struck. The final torpedo destroyed the engine room and the *Gustloff* lumbered with an increasing list until 70 minutes later she sank. In that time 1230 people somehow made it to safety, but the thick ice on deck had prevented the launch of many of the lifeboats. Others, too heavily laden, capsized as they were lowered into the sea. At least one was so full that its cables snapped. Another that had successfully made it to sea level was crushed when an anti-aircraft gun fell from its mounting as the great ship listed. The water was littered with dead children. Many had been given priority in the distribution of lifejackets, but they were designed for adults and would not support a child's head.

Suppressed History

One of the most remarkable facts about this story is how little known it is. Even at the time it went largely unreported. Russian newspapers made no mention of it at all. German newspapers were reluctant to worsen already tumbling morale. Only in one piece of British propaganda did it make headline news – *Nachrichten für die Truppe* ('News for Soldiers'), a leaflet dropped by bombers on pockets of resisting German troops.

Great uncertainty still surrounds the incident. Many of the survivors who have been interviewed by researchers have refused to have their names mentioned or their contributions defined; this is not exactly 'open' scholarship. But ever so slowly this profile is improving. A few books and articles have been published on the sinking, although the best of these are only available in German. A number of television programmes have also been made and more are promised. The ship herself has been found, explored and – predictably – looted. In 1990 Russian president Mikhail Gorbachev awarded Marinesko the title of 'Hero of the Soviet Union' and he has become something of a role model for the Russian submarine service. The submarine museum in St Petersburg is named after him and there are monuments to him at Kaliningrad (Königsberg), Kronstadt and Odessa. And David Frankfurter, the man who began this story by assassinating Wilhelm Gustloff, died in Israel in 1982 where he had lived for almost 40 years, having been released from prison at the end of the war.

USS *Indianapolis*

30 July 1945

The cruiser USS Indianapolis *was the last major warship to be lost in the Second World War. Just eight days after she was torpedoed by a Japanese submarine, an atomic bomb was dropped on Hiroshima, ushering in the end of the war. The* Indianapolis *had also played a vital role in the story of the world's first nuclear weapon, but she was to be remembered for the terrible plight of her surviving crew after the sinking, who spent days in the waters of the Pacific at the mercy of predatory sharks.*

On Monday 6 August, 1945 an American B-29 Superfortress dropped an atomic bomb on the Japanese city of Hiroshima. More than 118,000 people died instantly, plus a further 20,000 from the effects of radiation in the following weeks. The temperature at the epicentre of the explosion reached several million degrees centigrade and generated winds in excess of 500 miles per hour (800 km/h). Eight days later another A-bomb was dropped on Nagasaki, which led directly to the surrender of Japan and the end of the Second World War. These were two of the most decisive events in world history, but as always with such moments there are stories within stories, untold tales that link apparently unconnected events. It is those tales that throw the better-known historical facts into a broader context, and make them even more powerful. So it is with the familiar story of the bombing of Hiroshima; few know that it is intricately tied in with another story, that of the shipwreck of the USS *Indianapolis*. In fact, so closely are these two events linked that the bomb itself, when it was released from that American bomber, carried on its outer casing a handwritten message to the Japanese from the American air crew. The message read: '*This one is for the boys of the* Indianapolis'.

The *Indianapolis* was a Portland-class heavy cruiser. A magnificent ship of nearly 10,000 tons and over 190 metres (600 feet) long, she had a crew of 1199. She was launched in 1932 and by 1945 had enjoyed a remarkable career. Fortunate enough to have been at sea during the Japanese attack on Pearl Harbour, she was one of those ships immediately sent out to hunt for the Japanese task-force as the American Pacific Fleet lay burning in Hawaii. She then became the flagship of Admiral Spruance's 5th fleet in 1943, and two years later was involved in the crucial invasion of Okinawa. The island of Okinawa lies off the southern tip of Japan and in 1945 became strategically critical for the planned US invasion of Japan. Blessed with fine natural harbours and numerous airstrips, the island could act as a bridgehead for US forces to launch the invasion of Japan itself. At this stage of the war, Japanese airpower, once so formidable and manned with highly skilled pilots in beautifully engineered planes, had been devastated by a string of American naval victories.

Japan's response was to use warplanes for *kamikaze* ('divine wind') attacks. Loaded with high explosive, they were flown directly into enemy ships by young pilots with only the minimum of training. The Americans countered this new threat by keeping their valuable aircraft carriers behind a screen of destroyers, cruisers and escort ships, which bore the brunt of any *kamikaze* attack. It was this strategy that first nudged the *Indianapolis* towards a fate linked with the atomic bomb that would later fall on Hiroshima.

At the invasion of Okinawa, 225 American warships were damaged, many by *kamikazes*, and the *Indianapolis* was one of them. She was forced to return to the Mare Island Navy Yard in California. After a major refit, she found herself ready for active service at a time when the war was about to take a decisive turn, and the US had very few naval resources to spare. The battle for Okinawa was soaking up ships like a sponge, which meant that wherever the *Indianapolis* was sent she would go there alone, with no escort to protect her from Japanese surface ships, aircraft or submarines.

Top-Secret Cargo

In San Francisco her captain, Charles McVay (1898–1968), received secret orders. He was to load aboard a special cargo, and deliver it to the Pacific island of Tinian. He had no other knowledge of his mission at all until that cargo arrived under heavy escort, and only then could he ascertain what his eyes told him. The secret cargo came in two separate components. The first was a large timber packing crate. Unremarkable in itself, it was stowed on deck and given its own armed guard with a 9-metre (30-foot) exclusion zone around it. Anyone approaching the crate was to be shot. The second part of the cargo was a black canister, some 45 centimetres (18 in) long and with roughly the same diameter. This was even more curious. The canister was given its own cabin, was welded to the deck and the cabin door locked. The captain was issued specific instructions that if anything should happen to his ship, the canister was to be given its own life raft and set adrift. Only then were the crew to look to their own safety.

And so the *Indianapolis* set off from San Francisco into the Pacific to deliver her cargo, which by now the irrepressible inquisitiveness of her crew had decided contained

Hollywood film star Rita Hayworth's underwear. Some surmised that it was gold bullion, while those with a good strategic grasp of the current state of the war thought that it might contain a biological or chemical weapon of some sort. No one could have known that it was in fact the crucial components of an atomic bomb; the bomb casing was in the timber crate on deck, while in the mysterious canister lay the Uranium-235 fissile core of the weapon. This represented half of the entire stock of U-235 available in America at that time, and was worth an estimated $300 million. The first ever test-detonation of an atomic bomb had only happened ten days previously in the desert of New Mexico. The lowly crew of this American warship had no conception that a new age of warfare had been born, and that they were playing a crucial part in its inception.

The voyage was uneventful. The *Indianapolis* unloaded her cargo on 26 July and headed for Guam, the largest island in the Marianas, where she refuelled and rearmed. She was then given her sailing orders: she was to report to Leyte in the Philippines, where she was to join US forces massing for the planned invasion of Japan.

Loss of the *Indianapolis*

The route she was to take was not unusual, and although it was acknowledged that there was a potential risk of attack from Japanese submarines, this threat was not considered any higher than usual, and the *Indianapolis*, as a heavy cruiser, was considered capable of looking after herself. Fast, heavily armed ships that could have been used to escort her had been diverted to Okinawa where they formed a protective shield around the carriers and the troop ships; ships that were more valuable, and more vulnerable, than the *Indianapolis*.

Thus the *Indianapolis* found herself alone, in the middle of the Pacific, being tailed by the *I-58*, one of only six remaining operational Japanese submarines. The submarine's sonar operator had picked up 'clinks' from the *Indianapolis*' galley when the ship was 12½ miles (20 km) away. The Japanese submarine was commanded by Mochitsura Hashimoto (1909–2000), and as soon as he was able to identify his target as a large American warship, he fired six of his torpedoes and felt the explosions reverberate through the water as two of them struck home with devastating effect. The entire bow section of the *Indianapolis* was blown off by the first torpedo and the wounded ship continued to plough forwards into the sea, gaping like a basking shark scooping up plankton. The water pressure crushed bulkheads as she rapidly filled with water. Moments afterwards the second torpedo struck amidships, detonating a magazine and splitting open a fuel tank. That explosion nearly broke her in two and poured hundreds of gallons of fuel oil into the sea.

Some sailors abandoned ship immediately, leaping as

much as 30 metres (100 ft) from the listing ship into the oily sea. Others waited for specific orders from their officers, who had little grasp of the situation. Those on deck had no knowledge of the chaos in the engine room or of the water forcing its way deeper into the hull. The ship was too badly hit for any damage-control reports to be made. The ship's PA system had stopped working and only one of her two radio shacks was operational. The shack nearest the stern, the furthest away from the explosions, was the only one that could be made to work. Distress messages were sent giving the location of the sinking, for by now it was clear the ship was doomed. The *Indianapolis* was rising up into the air, her great propellers silhouetted against the sky, and the men cast into the fizzing, boiling sea. Within just 12 minutes of the first torpedo striking, she had completely disappeared and was on her final journey, an accelerated dive to the seabed, three and half miles (5.5 km) below, at roughly 11 metres (36 ft) per second.

Left for Dead

As many as 900 men made it into the sea, which is remarkable in itself given the scale of the explosions and the speed of the sinking. Roughly half that number had life jackets of one sort or another. Some were inflatable while others were made of cork. A small number of life rafts had also been thrown into the sea along with a handful of 'floater' nets – large floating nets onto which as many as 60 sailors could clamber. A huge sea of debris also began to appear as it was forced from the hull in pockets of water. Crates of supplies and bodies were blasted clear of the sea as they surfaced in air bubbles. Many in the sea were already terribly injured, with fractured bones, scalds and burns from the torpedo explosions. Others had hurt themselves jumping into the sea. One sailor leapt from the stern, only to fall 12 metres (40 ft) on to the ship's steel rudder. He could not move his legs. The torpedo had struck at night, so very few sailors had any clothes on at all apart from their underpants. But even the uninjured immediately began to suffer. It was impossible to avoid the slick of oil over 5 centimetres (2 in) thick that lay on the surface of the sea. It coated everything, and filled up their eyes, ears and mouths like treacle. It had been almost impossible not to inhale the acrid smoke from burning oil or the oil itself, and now, struggling to stay afloat, the first reaction of most of the sailors was to vomit compulsively and repeatedly.

The most badly injured died within hours, some within minutes of the ship going down. Their bodies were quickly robbed of their life jackets. A fortunate few, including the captain, found themselves floating near life rafts that had small boxes of supplies; tinned spam, crackers and malted milk tablets designed to slake thirst. None of the fresh water tanks that were issued for these rafts had been filled at Guam. In the early stages of their ordeal, those with cork life-vests fared well, but gradually the cork became waterlogged and the sailors' heavy heads dropped towards the sea. Those not strong enough to keep their chins up drowned. The inflatable life jackets were the worst. The seams that bound the tubes together rotted in the oil and the precious air leaked out. Sailors fought each other to get on rafts, and then fought to stay on them. Those in the sea huddled together, for their combined mass was sufficient support to keep their heads above water. Like this, they drifted in the vast expanse of the Pacific Ocean, 650 miles (1040 km) due west from Leyte and the same distance due east from Guam, clinging to the hope that their distress message had got through. The radio technician who had made the calls was confident, as was the captain, and those who were near these men were filled with a false hope.

The power of that hope, false though it was, kept those men alive as their ordeal began to take a sinister turn, for circling below the groups of survivors, sometimes nudging their feet, sometimes raising their snouts above water, was an ever-increasing shoal of sharks. Sailors know their fish. They recognized the dorsal fins of Blue sharks, Makos and Tiger sharks. The sharks had taken no time at all to arrive, but few sailors were surprised. It was well known in the service that sharks followed warships, perhaps attracted by the noise, vibration and electrical charge of the hull; certainly attracted by the vast quantities of sewage and waste produced by over 1000 men that is thrown overboard or pumped into the sea. The isolated dead bodies, floating in their life jackets, were the first to disappear. Some were pulled from beneath; while others were driven clear of the sea in a ferocious strike.

Saved at Last

Meanwhile, life in Leyte continued very much as before. Three radio operators received worrying signals, apparently sent by the *Indianapolis*, claiming that she had been torpedoed. But this was not necessarily unusual. It had long been Japanese practice to transmit disinformation on emergency channels to lure American forces into traps, divert them from their original intentions, or merely to test the strength and response time of American forces. It was such a well-known ruse that it had become official practice not to react to such a signal until its authenticity could be verified – which required sending a response to the signal, and receiving a satisfactory reply. Dutifully, this action was taken, but by the time a response was sent, the *Indianapolis* was lying at the bottom of the Pacific. Planes flew regularly over the survivors, but at altitudes of over 1500 metres (5000 feet), pilots and their crew could see little. Nor did they have any reason to be searching, and without specific instruction to do so, the task of flying absorbed all their attention. Unknown to them, therefore, the only hope for the survivors of the *Indianapolis* was for the port authority in Leyte to realize that they had not appeared at the scheduled time and to investigate. As it was, when the day of their expected arrival came and went, a simple note 'overdue' was put next to the name of the *Indianapolis* and no further action was taken. It was assumed that a last-minute alteration to her schedule had been made. To do so without informing the relevant authorities was rare indeed, but it was not unheard of. And in the case of the *Indianapolis,* a well-known personal favourite of Admiral Spruance, it was a distinct possibility. There matters rested for a further two days.

By now, conditions on the rafts and in the packs of floating survivors were deteriorating badly. Much later, Jack Milner, one of the *Indianapolis'* radio technicians, wrote: '*As the days dragged on I thought less and less as I dreamed more and more*'. He was not the only one, and their dreams were full of visions – the men were beginning to hallucinate from hunger and thirst, and in one group of survivors a vicious fight broke out. Some were drowned and others stabbed. During the day the sun burned unremittingly on their heads; they could even see the sun with their eyes shut. Those who had spare material made blindfolds. At night, temperatures dropped and brought on uncontrollable shivering. They put anything they could find in their mouths to try and stop their teeth chattering.

One sailor chewed right through a thick piece of rope. Those that drank salt water died horribly, some very quickly indeed; the human body is simply unable to process that much salt, and the brain and vital organs rapidly shut down. A body already dehydrated has little defence against the poisons entering the body. Hundreds had already died, and more were dying by the minute. Even the healthiest, those in the rafts with access to some food supplies, would be able to survive little more than another 24 hours.

Finally, it was the great oil slick that saved them. Spotted from a reconnaissance bomber out on a regulation patrol, the pilot lowered his altitude and tracked the slick, in the belief that it had come from a Japanese submarine. Instead, he saw groups of men dotted in the water, waving frantically. He marked their position by dropping bombs of orange dye into the water. He jettisoned the survival equipment he had on board, and reported his position back to Leyte. At roughly the same time, those who had been expecting the *Indianapolis* to arrive in Leyte were beginning to question their initial assumptions, and very quickly indeed the pieces fell into place. Horrified, the naval authorities on Leyte organized a rescue operation.

Vilification and Restitution

We do not know exactly how many men were pitched into the water as the *Indianapolis* went down. What is certain is that only 316 survived their ordeal in the water. While the immediate culpability for the disaster lay with the Japanese sailors who torpedoed her, it was a legal act of war and seems to have been respected as such. The bigger question, and the one which has made this one of the longest running sagas in American naval history, is how the United States Navy managed to leave almost 1000 American sailors to fend for themselves in the Pacific for almost five days. The official response, which won the navy no friends at all, was to point the finger firmly at Captain McVay, who was duly court-martialled.

The captain was responsible, they argued, because he had failed to follow a zig-zag course, the standard submarine-evasion technique, and had failed to transmit a distress signal. From the limited evidence available then, the case was clear-cut, and McVay was reprimanded and demoted. In doing so they made him the first US captain in history to be court-martialled and found guilty for losing his ship as a result of an act of war. Worse still, he became the public face of the loss of the *Indianapolis* and received hate mail from relatives of his crew for the rest of his life. Mothers blamed him personally for the deaths of their sons; wives for the deaths of their husbands; and many years later, sons and daughters for the deaths of their fathers. Although McVay knew in his own heart that he had done all that he could on that night, he could never escape the wreck of the *Indianapolis*. This man, who had survived four and a half days of hell, in which we know from his own hand that he had prayed and begged to God to survive, was so haunted by the wreck in later life that on the morning of 6 November, 1968, he shot himself with his .38 service-issue revolver at his farm in Connecticut, aged 70.

The trauma of those days at sea greatly affected most of the *Indianapolis'* survivors in later years, and it is one of the few examples of a shipwreck from which we can gain some idea of the haunting impression that such a disaster can have on the mind. The ship's doctor was unable to recite the Lord's Prayer again without breaking into tears, so regularly had he repeated it at sea. He had to avoid going to church for the rest of his life. Another survivor is restless without a glass of iced water at his side, so fearful is he of going thirsty. Many others desperately swim from sharks in their dreams. At the time of writing, there are 38 living survivors.

Those survivors have met regularly since the first reunion in July 1960. That day, they sat together in designated seats, recreating the groups in which they had floated while waiting to be rescued. It was only the start of the healing process for them. Deeply buried anger, pain, shame and confusion poured out as they tried to make sense of what had happened to them, and why. One of their greatest concerns was the fate of their captain. To a man, they stood by Charles McVay, and when he arrived at the first meeting they had lined up to salute him, tears streaming down their faces. The meetings themselves generated public and political interest, and so too did the steady trickle of books that were published about the tragedy. These too had their own power; the latest book not only describes how the men's physical and mental conditions deteriorated, but it explains from a medical perspective why it happened and it explains why some were stricken by madness and others not. It was one of the most important discoveries for many of the survivors. They could now rationalize their behaviour in a way that had previously been impossible. Those who felt shame for their actions could accept and understand their behaviour, and in part at least, free themselves from the grip of this awful memory. With this tide of public interest came increased political pressure to reverse the conviction of Captain McVay. In 2000 Congress duly passed a resolution exonerating him from responsibility for the loss of the *Indianapolis*, and in 2001 the navy officially cleared his professional record. The story of the *Indianapolis* is important on so many levels, but it is perhaps this demonstration of the ability of history to help us understand our past, and to help us right past wrongs, that rings the loudest.

Nuclear Power, Oil and the Atomic Bomb

By the end of the Second World War, two of the greatest modern aids to seafaring had become highly developed: RADAR and SONAR. Marconi knew from the 1930s that radio waves could be used to detect remote objects, but it was a number of years before RADAR become operational, and it was first used by the British in the Second World War to forewarn of German air-attack. By 1946 RADAR was powerful enough to measure the distance to the moon as 240,000 miles (384,000 km). It can be used to determine the presence and range of an object, its position in space, size and shape, velocity and direction of motion, and it can do so in the darkest of nights or the thickest of fogs.

For mariners, these technologies meant that they could easily measure their distance from the nearest coastline, and they could even identify it from the distinctive pattern that it produced on the RADAR screen. Nearby vessels could be tracked in their course and speed to avoid collision. At anchor in a storm RADAR can be used to monitor a ship's position closely, to see if she is dragging her anchors. It can even identify approaching squalls. SONAR works on the same principle, but it uses sound waves. One of its most valuable applications is the echo sounder, which continually measures the depth of the sea underneath a ship by emitting a signal and measuring the time taken for it to return. SONAR also has myriad applications in military vessels through its ability to locate and track underwater objects such as submarines, mines, or even hostile frogmen.

Both SONAR and RADAR are exceptional safety tools for the modern mariner, but they have also been used to improve modern weaponry. Because SONAR and RADAR can 'see', they can be used to guide weapons to their target. Targeting ships with traditional non-guided weaponry, either from the air, the surface or submerged, is a very tricky business indeed. Not only does the target move along her course at an unpredictable velocity and on an unpredictable course, she also rolls with the motion of the sea. So too does the ship that is doing the firing. Moreover, when an attack is launched over a considerable distance, the time delay between launch and impact is sufficient for a degree of prediction to be necessary in the course, speed and motion of the target. With weaponry guided by SONAR, RADAR, heat or satellite, the likelihood of a successful attack is vastly improved.

Nuclear Submarines

Improvements in the accuracy of weaponry were also matched by great advances in their power. The bombing of Hiroshima and Nagasaki in the final weeks of the Second World War focused military scientists on the adoption of nuclear power for military use and it did not take long for the atomic fission bomb of the 1940s to be replaced by the far more destructive hydrogen fusion bomb of the 1950s. Intercontinental ballistic missiles were soon developed to launch these missiles over enormous distances, and they were carried in the largest, fastest, most heavily protected, and crucially the quietest submarines ever built.

These submarines were also the focus of the second significant application of nuclear technology: nuclear powered propulsion. Traditional submarines could only operate submerged for a limited period before they were forced to surface and recharge their batteries. Engines powered by nuclear reactors do not suffer such restrictions; these vessels can travel hundreds of thousands of miles before the nuclear core, which produces the energy required to power the engines, has to be replaced. Now, submarines with devastating and accurate firepower could submerge and simply vanish. The first nuclear submarine was the American USS *Nautilus* of 1954, and in 1958 she passed under the North Pole. That same year another American submarine, the *Skate*, actually surfaced at the pole.

Not only could a submarine navigate its way exactly to the North Pole under ice, but by passing underneath the polar ice-pack, it could take a shortcut from one side of the world to the other; submarines could easily and quickly enter the Pacific from the North Atlantic. And if they did not want to go that far, they could enter the Barents Sea, a perfect location from which to target the industrial heartland of Russia. The equivalent would be for a Russian submarine to launch an attack on the great inland industrial centres of the US from Canada's Hudson Bay. The Arctic ice-pack, moreover, is an almost perfect environment in which a submarine can remain elusive. The permanent noise of grating bergs and plates of ice makes submarines near the surface almost impossible to detect with SONAR, while the ice itself shields them from the prying eyes of aircraft.

When the Russians followed the American *Nautilus* in 1959 with their first nuclear submarine, the ill-fated *K-19* ('Widowmaker'), the Cold War began to reach unprecedented levels of tension. Soon the commander-in-chief of the Soviet navy bragged that they had so many submarines that they could assign some of them to catch herring in the North Sea. By 1970 the combined Atlantic and Pacific Russian submarine fleet exceeded 370.

Modern Navigational Aids

The constant drive for technology to stay in the race of the Cold War led to many other important technological developments. That which is best known to mariners today is GPS (Global Positions System), which uses satellite signals to pinpoint location. A network of 24 Navstar satellites orbits the earth every 12 hours in a formation that ensures that anywhere on earth is always in radio contact with at least four satellites. The system reached its full operational capability of 24 satellites in 1993, and since then the most sophisticated GPS systems have been accurate to within a matter of feet anywhere in the world, at any time of night or day, and in any weather.

GPS is actually based on a much older global navigational system run by the Americans called OMEGA. First operational in 1968, but abolished with the advent of GPS, OMEGA was a global network of radio signals beamed from eight locations. In America there were two stations, in North Dakota and Hawaii; the others were in Argentina, Norway, Liberia, France, Japan and Australia. The radio towers were engineering feats themselves: the Norway station claimed the longest antenna span in Europe, the Japan tower was the highest in Japan, and the

Argentina and Liberia towers were the tallest structures on their continents. Two or more of these signals could then be used to fix one's position to within 4 miles (6.4 km) on any ocean in the world.

GPS works on exactly the same principle, and both rely on their functioning by accurately measuring the time difference between the signals received. Because in GPS the distances involved are so vast, this was only possible after the invention of the atomic clock, which uses the ultra-reliable resonance of the atom to measure time, and is accurate to a billionth of a second. Every single satellite carries four such atomic clocks, and in each broadcast it emits its location and the exact time: at the heart of navigation, the challenge of telling the time at sea is as important now as it was when Harrison solved the problem of longitude in 1760. Now GPS is used to navigate, to guide weaponry and to locate stricken ships and casualties.

The Versatile Helicopter

Another modern invention which is crucial to the history of shipwreck, but which so rarely features in it, is the helicopter. The earliest pioneers created working helicopters in the 1930s, but they were not mass-produced until the 1940s, and only saw their first role in military operations in the Vietnam War. Their ability to launch vertically gives them extraordinary flexibility, and by the 1970s any warship of a reasonable size could accommodate a helicopter. They are easily armed with light guns and guided missiles to target ships or submarines, and are particularly useful for carrying 'dipping' SONAR buoys which can be submerged wherever desired to hunt a submarine. They also play a key role in search-and-rescue operations, from ships or land-based coastguard stations. Able to hover, they can winch survivors from stricken ships, life rafts or the sea.

In almost all such rescue operations, helicopters are used in integrated operations alongside modern lifeboats, which, owing to significant innovations in marine architecture in the second half of the 20th century, can venture to sea in almost any conditions. Modern lifeboats can still be capsized, but they are now self-righting; they remain afloat even when they are upside down, and possess positive righting levers throughout the full 360 degrees of heel. The first of these came into service in 1954, and it used an ingenious self-filling and draining water-ballast tank located immediately under the deck, into which ballast was transferred as the boat rolled over. The offset weight of the transferred ballast, together with asymmetry in the engine casing, provided the self-righting capability.

Age of the Supertanker

Many of the ships that suffered shipwreck from the 1950s had also changed dramatically, principally in their size.

These are the years in which container ships and supertankers first appeared. It was discovered that once a certain size had been reached, the capital cost of a tanker rises very slowly: the cost of pipes and pipelines to feed the tanks varies little, as does the crew requirement for wages, provisions and accommodation. In short, the larger a cargo ship is, the more economic it is. In 1959 the Japanese launched the *Universe Apollo*, the first tanker larger than 100,000 tons. Only seven years later, the *Idemitsu Maru* was launched, more than twice the size of the *Universe Apollo*. In 1970, tankers of over 500,000 tons plied the oceans, their bellies full of oil. Driven by an insatiable global thirst for oil, these tankers make up a major portion of the world's shipping. It is hardly surprising that the most disastrous modern shipwrecks involve oil tankers. For example, in March 1978 the tanker *Amoco Cadiz* ran aground off Brittany and bled 260 million litres of oil into the Atlantic. In December 1987, 4300 people died when the Philippine ferry *Dona Paz* collided with the tanker *Vector*, while in 1989 the *Exxon Valdez* ran aground off Alaska and ruptured her hull. Clean-up operations on the beautiful coastline of Prince William Sound cost her parent company $2.1 billion and a further $1 billion in lawsuits.

Such wrecks seem particularly dramatic because for many of us they occurred within living memory and in an age where pictures show a tragedy unfolding more effectively as words describe it, but it is now possible at least to locate such disasters in a broad and detailed historical context. In recent years our knowledge of the development of seafaring and of shipwreck has started to satisfy our interest in it. This change largely came about because of the invention of the aqualung in 1943, which opened a new world to human enquiry. For the first time, the sea itself could be studied in detail by non-specialist divers, and the wrecks that litter its floor used to illuminate the past.

The *Nagato*

29 July 1946

In a demonstration of the fearsome destructive power of the atomic bomb, in 1946 the United States conducted Operation Crossroads – the sinking by nuclear explosion of a large target fleet at Bikini Atoll in the South Pacific. This ghost fleet of decommissioned vessels included several proud American veterans of the Pacific War, such as the obsolete carrier USS Saratoga. *But more significant was the presence of the flagship of the Japanese Admiral Isoroku Yamamoto – the ageing and crippled battleship HIJMS* Nagato. *Her destruction delivered the final symbolic* coup de grâce *to the once-mighty Imperial Japanese Navy.*

John Hersey (1914–93) was an exceptionally talented journalist. Born in China, where he lived until the age of ten, he also spent a great deal of the Second World War deep in the Pacific, reporting on the progress of the conflict for a variety of American newspapers and magazines.

Hersey's familiarity with the Far East made him the obvious choice for the editor of the *New Yorker* to commission to travel to Hiroshima in May 1946 to talk to witnesses and describe the effects of the atomic bomb. His story caused a sensation. The resulting special edition of the magazine, which contained nothing other than Hersey's account, sold out within a few hours of its publication. Newspapers and magazines from all over the world queued up to serialize his story. Penguin Books published it in the UK three months later. Today Hersey's story is still powerful. It follows six survivors of that day: a German Roman Catholic priest, a Red Cross hospital doctor, a doctor with a private practice, an office girl, a Protestant clergyman and a tailor's widow. The account describes in detail the physical and emotional effects that the bomb had on these people, their friends, family, and on Hiroshima, their home.

The Awesome Power of the A-bomb

Very shortly after Hersey's *Hiroshima* was published, the first photographs of an atomic explosion also began to appear. These had not been taken at Hiroshima, Nagasaki or even in the desert of New Mexico where the one and only atomic test prior to the bombings had taken place, but at an obscure group of tiny islands, deep in the Pacific, known collectively as Bikini Atoll. They were photographs of two test explosions of atomic bombs, carried out by the Americans. In the late summer of 1946, therefore, the shocking power of the atom bomb, witnessed in both pictures and words, became popular knowledge for the very first time.

Those who were granted access to the test results were able to form a detailed and scientific picture of the effects of an atomic bomb detonated in the air compared to one detonated underwater. Moreover, for the very first time it was possible to examine the effects of an atomic explosion on a fleet of warships, for at the epicentre of those blasts at Bikini Atoll 77 target ships were anchored, laden with test equipment and animals.

This was the real crux of the Bikini tests. The bombing of Hiroshima and Nagasaki had demonstrated the force of an airburst against a land target, but it is easy to forget that beyond the obvious destruction caused, very little indeed was known about the effects of the atomic bomb. Only three had ever been exploded, and two of those had been in anger, with no scientific observation possible. Scientists and medics were too late when they arrived in Japan to assess the cause of the horrific injuries they saw: were the deformities and scars caused by the initial blast, by debris or by later radiation? Moreover, Hiroshima and Nagasaki were both ports, but both bombs were exploded around 2745 metres (3000 yards) from their harbours, and as yet no valuable data had ever been gathered concerning the effects of a nuclear explosion on ships. This was a major concern to every navy in the world. Newspaper headlines immediately after the attacks on Japan were unsurprisingly sensationalist, and more than one journalist declared that the ability to wield such power would bring an end to war itself. Others claimed that naval power had become obsolete.

Military analysts were more circumspect, but undoubtedly the bombing of Hiroshima and Nagasaki had raised many more questions than it had answered. How would this new weapon be used in the future? How could it be defended against? Was there any future role for conventional warfare, and if so what form might it take? Would navies still retain their value in defending maritime trade-routes, or would they simply become obsolete and irrelevant if the ports at which the trade was to land were obliterated? How easy would it be to target a fleet of warships with an atomic bomb, and what would its effects be? Some believed that the steel hulls would simply disintegrate if hit by a nuclear explosion detonated underwater.

These were some of the questions that the US Navy set out to answer at Bikini Atoll in July 1946. Not only would it demonstrate to the world its commitment to nuclear warfare, but it would do so by destroying a fleet of target ships so large that it ranked as the sixth most powerful naval force in the world at that time, such was the power, strength and status of America in those years. This was a calculated and carefully stage-managed demonstration of that power on an unprecedented scale; the first explosion was witnessed by 114 American reporters. Not only was it an opportunity to showcase American might, however, but it was also an opportunity to impress upon the world her dominance over her recently vanquished enemy. Of all the ships that were taken to Bikini Atoll, the one placed closest to the underwater blast; the one that would be guaranteed to be destroyed; the one that carried with it very little test equipment and was largely ignored in the later analysis of the results was the Japanese battleship HIJMS ('His Imperial Japanese Majesty's Ship') *Nagato*, former flagship of Admiral Isoroku Yamamoto (1884–1943), architect of the Pearl Harbour attack.

Sinking the 'Nasty Naggy'

Built in 1917 to rival the mighty British battleships of the *Queen Elizabeth* class, the *Nagato* was a colossal ship of 32,730 tons, armed with 16-inch guns and was a veteran of nearly every major conflict with the American navy in the Pacific War. By 1946 she was the only one of the ten Japanese battleships that had been afloat in 1941 still in existence. She was considered a particularly ugly ship by the Americans who called her the 'Nasty Naggy'. The Director of Ship Material at Bikini described her as a '*massive and brutal structure, having neither grace nor beauty.*

She gave no comforting illusion that war is anything but an ugly, brutal and totally insane occupation of man.' Her destruction would finally extinguish the last remains, both physical and symbolic, of the once mighty Imperial Japanese Navy, and the Americans saw no more fitting way than to submit her to the same fate as the cities of Hiroshima and Nagasaki. The *Nagato* was consequently summoned from Tokyo where she lay, badly damaged from the American bombing at the end of the war, and still with a large number, perhaps as many as 100, bloated Japanese corpses inside her flooded compartments. When she arrived at Bikini she had to be permanently pumped out, so near was she to sinking. But considerable resources were expended to keep her afloat, until, almost on life-support, she was anchored within feet of the coming atomic blast.

The first test, on 1 July, was known as 'Able' and was the detonation of a 21-kiloton atom bomb 158 metres (518 ft) above the surface of the lagoon. It was witnessed by thousands of sailors on support craft many miles away, who were instructed to turn their backs on the explosion and to cover their eyes. The sailors claimed they could see the bones of their arms through their closed eyelids. The blast of air screamed at over a mile per second among the fleet, followed by a colossal fireball. Several miles away an army observer watching the aftermath of the blast through a telescope received a black eye as it slammed into his face. The results were, to many observers, surprising.

Only five ships had actually been sunk, and the *Nagato* had survived. Six others had been severely damaged and many were on fire, but the vast majority of the ships themselves were still much as they had been, if severely scorched, twisted and bent. Only one of the target ships' supply of munitions had exploded despite the strength and heat of the fireball. But although many were still afloat, those who visited the ships shortly after the blast were well aware that it would have been almost impossible for any of those ships' crews to have survived. Vast numbers of the test animals used for the Able tests died in the blast or soon afterwards from its effects.

The second test, on 25 July, known as 'Baker', was designed to demonstrate the effects of an atomic bomb detonated underwater. The results, this time, were more spectacular. A column of water almost 300 metres (975 ft) thick rose at a speed of 762 metres per second (2500 ft/sec), its centre of super-heated steam rising almost four and a half times faster, at 3350 metres per second (11,000 ft/sec). A minute after the explosion, the column of water was 2316 metres (7600 ft) high. The shock wave underwater ruptured bulkheads while on the surface, tidal waves almost 30 metres (94 ft) high travelling at 45 knots engulfed the ships. When the column of water, an estimated 2 million tons, finally fell back to earth, the radioactive spray covered every surface and penetrated every compartment. As the mist cleared, the *Nagato* was seen sinking slowly by the stern, one of nine ships to have been fatally damaged by the blast. She took four days to finally capsize and sink.

Dawn of the Cold War

In 1989 divers from the American National Parks Service went back to Bikini to survey the wrecks there. Each had been sunk with oil in its fuel tanks and ammunition in its magazine; these wrecks were clearly dangerous to both the public and the environment, and yet almost nothing was known about what was actually there, or how it was changing. The *Nagato* was discovered on the sea floor, capsized by the weight of her control tower, almost 30 metres (100 ft) tall. She descended at a sharp angle and then buried her stern in the mud. The bows gradually came to rest and the hull tore itself in two under its own weight. There, in the depths of Bikini Atoll, the *Nagato* is both a monument to the great Japanese successes and failures of the 1940s, and also testament to the dawn of a new era of warfare.

By demonstrating, again, the shocking power of destruction of the atomic bomb, America had made one thing very clear: war between two nuclear powers could not be risked. Military victory through brute force would never enjoy the same status that it had done since the beginning of time. Now, although conventional warfare would still have its place, relations between belligerent nuclear powers would be characterized by shows of force and resolve, by secrecy, bluff and counter-bluff. This was an era in which hugely powerful nations would flex their muscles in countries far distant from their shores, with no direct threat to their own security. The Soviet Union would take its stand in Afghanistan, and America in Vietnam. These examples of conventional military force, allied with unassociated displays of nuclear capability, would characterize the conflict between nuclear powers.

The Bikini tests also demonstrated that the role of navies would not be as drastically altered as many feared. Naval bases where ships were moored in compact groups were clearly at risk from nuclear attack, but ships at sea, dispersed in convoy, were not so. Submerged submarines, although very vulnerable to a submerged detonation, were also extremely difficult to target accurately from aircraft. Britain's commitment to a naval nuclear arms programme remained circumspect. The British believed that the power of an atomic bomb was expended very uneconomically against a warship or a group of warships. One official report noted soberly: '*The loss of one capital ship was nothing if compared to the loss of an ordnance factory specializing in guns, together with 10,000 industrial workers and their families.*' The immediate British response, therefore, was to alter the practices of the navy so as to increase their defence against such an attack. Ships would spend as little time in harbour as possible; they would be replenished with both fuel and supplies at sea. They would be dispersed among ports worldwide, and even within those ports themselves. Meanwhile, British adoption of naval nuclear weapons was tentative.

The Americans, on the other hand, actively and immediately developed naval nuclear weapons to be used offensively. Nuclear depth charges, nuclear missiles, nuclear torpedoes and nuclear shells were all designed and built. In fact the United States pursued the development of all types of nuclear weapon wholeheartedly and by 1986 had manufactured at least 60,000 nuclear warheads of 71 types for 116 different weapons. By the late 1950s, fast, nuclear-powered American, British and Russian submarines could spend months submerged, and deliver a nuclear warhead anywhere in the world. That relationship between the world's navies and atomic weapons, now so influential in international relations, can be traced right back to Bikini Atoll, and more specifically, to the *Nagato*, the first warship sunk in anger by a nuclear weapon.

The *Torrey Canyon*

18 March 1967

Even after many more recent high-profile oil spillages involving terrible damage to marine life, the wreck of the Torrey Canyon *remains one of the most significant tanker wrecks of all time. One of a new generation of huge supertankers, her bulk and lack of manoeuvrability were to prove fatal when she found herself in harm's way off the Isles of Scilly in March 1967. Navigational incompetence, inexperience on the part of officials and inadequate emergency procedures all contributed to an environmental disaster that yielded many important lessons for the future.*

The oil tanker *Torrey Canyon* was named after an oilfield 75 miles (120 km) northwest of Los Angeles, which in turn was named after John Torrey (1796–1873), a physician, botanist and geologist who was chief assayer of the United States Assay office for 20 years. In the spring of 1967, she was carrying Kuwaiti oil to the refinery at

Milford Haven in Wales. She was an unremarkable ship, named after a man no one in Europe and very few in America had ever heard of, making an unremarkable voyage. Until, that is, the morning of 18 March, when the ship's chief officer was amazed to see the Isles of Scilly to port; by his reckoning they should have been to starboard. The error had been caused by ignorance of an easterly current; neither he nor the captain had read the sailing directions. The chief officer altered course, but was overruled by the captain once he discovered their real position. By turning them back onto their original course, however, the ship was now headed straight for the Seven Stones Reef. Her captain now had a decision to make – should he go around the Scillies, leaving them to starboard, or between the Scillies and the Cornish coast? He knew that the latter would save him a great deal of time, and so he chose to keep the Scillies to port. Not only that, but he decided to go between the Scillies and the reef, through a very narrow channel.

It was not a good decision. The passage was populated by fishing boats, which, because of their heavy nets, retain right of way at sea: any other vessel, however big, must go around them. Moreover, the *Torrey Canyon* was one of the largest afloat, and as such, was unable to alter course quickly and precisely. When her captain realized he was too close to the rocks, he was prevented from turning in his desired direction by a fishing boat, which had hoisted warning flags. The tanker was too cumbersome to make her way to safety in the opposite direction. Moments later she struck Pollard's Rock, only 16 miles (26 km) off the British coast, and stuck fast. Only 11 ships of the 150 that are recorded to have struck Pollard's Rock have been successfully refloated. The granite of which it is made is sharp and harder than steel. With the relentless momentum of the tanker's weight, the rocks tore open her hull. Within hours she was listing to starboard and her engine rooms were flooded.

The Disaster Unfolds

At a little under 300 metres (974 ft) long, the *Torrey Canyon* was the largest ship in the world ever to be wrecked – which in 1967 was newsworthy in itself – yet the truly striking thing about this ship was that she carried in her hold 119,328 tons of crude oil, and immediately after she struck Pollard's Rock, it began to bleed from her tanks into the sea. Until then the largest oil spill had been perhaps 10,000 tons; not enough to make headline news, but easily enough to alert the authorities to the potential environmental, economic and political damage of a major oil spill. Nevertheless, any research undertaken or procedures put in place to deal with oil spills were all based on theory. No scientist or politician had ever had to deal with such a catastrophe in practice. There was, moreover, no legal infrastructure that dealt with oil pollution; there was simply no precedent. Nor was there a government department with the specific responsibility of dealing with oil pollution: the British ministries of local government, technology, transport, agriculture and defence all had what they termed a 'confused interest' in the problem.

The most immediate of those problems was what to do with the ship. She was in international waters and the oil could not be destroyed until the owners gave their consent. A Dutch salvage company had also claimed the rights of salvage, and there was a hefty profit in it for them if they could refloat her. The British government, under advice from the Royal Navy, wanted to set fire to the ship where she lay. The underwriters of Lloyd's did not want to see such a cargo simply destroyed. The voices of the local fishermen or the owners of beachside hotels and campsites in Cornwall were distant indeed. With no precedent to consult, inaction prevailed in those first crucial hours.

Eventually the operation began. The oil would not be burned and the vessel destroyed as the navy recommended – rather, the sea was sprayed with detergent by a fleet of tugs and naval vessels. Journalists who watched the efforts of one spray-ship from an aircraft 150 metres (500 ft) up said it seemed futile, and that the foam was like a speck on a long, broad, brown floor. The slick was then 20 miles (32 km) and 5 miles (8 km) wide. It would hit the

Cornish coast within three days. The salvors, meanwhile, tried to refloat the ship which was becoming increasingly unstable as the swell worked her hull back and forth against the stones, and the volatile gases emitted by the oil collected in pockets of the hull.

Three days after the *Torrey Canyon* struck, a huge explosion blasted two salvors off her decks and into the sea. They were both rescued, but with extreme difficulty as the oil in the sea coated their skin and clothes and made them very hard to grip. One of the men died from the wounds suffered in the explosion, with blast injuries to the lungs and brain, and massive internal bleeding. The tanker was temporarily evacuated, but the Dutch salvage team, having invested so much money already, and having lost one of their crew, reboarded. Their chances of refloating her, they estimated, still stood at 50/50. Even if they could bring back a section of her stern with its valuable machinery, the salvage reward would still make the enterprise extremely worthwhile.

Thus the situation stood on Wednesday 22 March. The slick approached the Cornish coasts while local businessmen and fishermen waited in horror and raged in disbelief at the government's lack of response. The navy was desperate to bomb the stricken ship while they relentlessly sprayed the slick with detergent, a thankless and largely symbolic task. With every hour that passed, more oil leaked out and the environmental threat became worse.

The government realized that its handling of the situation would define their administration; the underwriters of Lloyd's were set to lose a vast amount of money while the Cornish locals and fishermen were set to lose, for an unknown period of time, their livelihood; and at the centre of it all a handful of Dutch salvors were in constant risk of their lives. It was a multi-faceted and extremely tense drama, played out in real time in the press, and the world was gripped.

Drastic Measures

It was at this stage that the weather took a turn for the worse. A force 7 gale pounded the steel hull again and again onto the razor-sharp reef until she broke her back and split in two. Her oil, which until now had been leaking from her hull, flooded out. There was now no hope of salvage and the environmental threat had escalated alarmingly; it was the worst possible outcome for everyone involved. On the very day that she broke up, scientists in Britain conducted emergency experiments to ascertain the best way to remove the oil. They poured 1000 gallons of crude oil onto a pond in a research centre in Sussex, simulated wind and waves and set fire to it. It burned very well, leaving only a fraction behind. The result was sent to the government, who instructed the Royal Navy Fleet Air Arm to scramble eight Blackburn Buccaneer strike aircraft loaded with 40 1000lb bombs. They were closely followed by Royal Air Force Hawker Hunter fighter jets, their spare under-wing fuel tanks lapping with kerosene. Once the first wave of planes had dropped their bombs and started a fire that poured thick black smoke almost 2500 metres (8000 ft) into the sky, the second wave opened their fuel tanks and drenched the flames in kerosene.

A formidable strike by any standards, but they were soon to discover that burning oil that has been spilled at sea presents complex problems. Oil released on the surface rapidly changes its properties as water-soluble toxic substances are released. The oil becomes virtually non-toxic, and the lighter and more volatile components evaporate so that the residues become progressively more difficult to burn. Moreover, if the oil can be ignited, it is very difficult to sustain burning as the heat of the fire is transferred to the underlying water, which quickly decreases the temperature of the oil below the required flash point. This the navy quickly discovered as only four hours after their initial strike the flames had gone out. They would not be so easily beaten, however, and ordered another strike, but this time the planes were armed with napalm and rockets. They were followed by more 1,000lb bombs, and through a sustained and ferocious attack that used in total over 160 1,000lb bombs, 11,000 gallons of kerosene, 3000 gallons of napalm and 16 rockets, the oil in the immediate vicinity of the wreck was destroyed. By then, however, oil slicks almost completely surrounded Britain's summer playground, the southwest coastline. That Easter weekend (25–26 March) the following announcement appeared in the 'Shipping casualties' section of *Lloyd's List*:

> *St Ives Bay: Oil coming in. Heavy pollution at Porthmeor beach. Harbour clean.*
>
> *North coast between Sennen and St Ives: Cape Cornwall very badly affected and more oil along the coast. No action being taken yet as more oil coming in.*
>
> *Sennen and Whitesands: Heavy Pollution. No work being done as more oil coming in.*
>
> *St Just: heavy pollution. With thick oil. No action taken as more oil coming in.*
>
> *Mousehole, Newlyn and Penzance: Clear.*
>
> *Porthleven: About half clear.*
>
> *Penberth, Porthgwarra, Porthcurno: Clear.*

Those areas described as 'clear', however, were only clear because the oil had not yet reached them. From Hartland Point in north Cornwall to Start Point in south Devon the foul tide slowly, but inexorably, came in.

Viscous Black Tide

The arrival of the slick off the coastline was no surprise; in fact, its course was precisely predicted. Oil, we know, travels on the sea at 3.4 percent of the wind speed, and as long as the wind is fairly strong, tidal currents have little effect on its progress. The progress of the slicks was tracked on computer, by sea and air. Most people had little idea of what to expect, but within hours the acrid diesel-fuel smell of the oil, the consistency of heavy engine oil and the colour of dark chocolate, would be overwhelming. The army was mobilized to fight the oil on the beaches while a tiny number of ships continued to spray the slicks with detergent.

When it finally came ashore the beaches were bulldozed and sprayed with highly toxic detergent that killed shellfish, particularly the limpets on the oil-covered rocks. This use of detergent was highly controversial. The detergent was well known to be lethal to many forms of life, but its potential impact, particularly on the valuable fish stocks off the Cornish coast, was almost unknown; it was simply 'hoped' that its effects would not be too catastrophic. In the fact that it was used so liberally to clean the beaches and to spray the slicks, the local fisherman saw preferment for the tourist trade over the fishing industry, which became an enduring source of conflict in local communities.

The authorities certainly did not hold back. The tiny and beautiful harbour of Porthleven on the south coast of Cornwall was doused with 29,000 gallons of detergent. Around half a million gallons of detergent were used in the entire operation. Now consider this: one part detergent in ten million parts of sea water is enough to kill the larvae of oysters and clams. At Porthleven, divers found legless and clawless crabs; the sea anemones completely disappeared; the tiny number of limpets left had lost their protective shells. At Fistral beach in Newquay, all of the red and green algae died, along with all the limpets and about 50 percent of the mussels and barnacles. On every beach sprayed with detergent there was only very scant evidence of normally abundant winkles, top-shells or dog-whelks, crabs and shrimps. Dark green, red and brown seaweed was bleached white.

The oil was also lethal to seabirds and other mammals. It clogged the wings of birds, preventing them from swimming and flying, and therefore from feeding, and when they attempted to clean their wings, the oil they swallowed poisoned them from the inside. Although many novice helpers became highly adept at catching and cleaning them, tens of thousands of seabirds still died. March in the Western Approaches is a very busy time for seabirds. Gannets, shearwaters, guillemots and razorbills converge in vast numbers as their migration comes to an end off the Isles of Scilly, which is also home to the much rarer Westcountry puffin. These birds, moreover, spend much of their time actually on the sea, and their colonies were devastated.

The Isles of Scilly remained remarkably unaffected, but one huge slick was blown ashore on the north coast of Brittany, where the local economy was driven by their famous oyster and mussel beds. The use of detergent, therefore, was not an option, and the oil approaching their coastline had been at sea for too long to burn. The fragmented nature of the north Breton coastline, moreover, meant that some beaches were inaccessible, and in some areas there was still a thick coating of oil 15 months after the wreck. As in Britain, ignorance of the impact of oil on the fish led to much concern among the fishermen, and more importantly, among the consumers. The French fish market collapsed.

Lessons Learned

These immediate effects of the oil spillage were indeed shocking, but of much more lasting significance was the global response to the disaster. This was, after all, the first major oil spill that the world had ever experienced. It led to important advances in science, law and environmental protection to deal with such a spill. One of the most remarkable advances in knowledge is that crude oil is one of the most complex mixtures of natural products on

Earth. So much so, that every oil has its own compositional features, which are typical and persistent – like a fingerprint. There was little doubt in the case of the *Torrey Canyon* where the oil came from, but more often than not, oil slicks are anonymous. It is now possible to link an oil slick to an oil field, or even a particular vessel.

The impact on the environment was without doubt the most shocking aspect of the *Torrey Canyon* disaster. At first, the British press published images of the wreck, its size being the newsworthy item. At the same time the government made a nominal donation of £1000 to fight the oil slick. But as the reality of the situation became increasingly apparent, so too did the public response become strong. The British government ultimately spent over £2 million dealing with the oil, and volunteers came out in their thousands to clean the beaches and treat the birds. It brought the world's attention to the enormous size that tankers had by then achieved, and the associated risks of transporting vast cargoes of oil and chemicals by sea. Complex questions arose, one of which concerned the risk inherent in transporting goods with large ships, rather than numerous small ones. Is a large ship more likely to be stranded than a small one? Are two 200,000 vessels more likely to collide than 40 15,000-ton ships? How can a government protect its shorelines from these great ships, in international waters, but so close to a nation's shoreline?

At a more basic level, the general public learned a great

deal too. A large number of books covering the wreck from various angles were swiftly published. One of those included an appendix on how to treat oiled birds. It offered the sage advice that one should '*make sure you don't get its beak near your eyes*'. Indeed. If one's bird has recovered, the manual is also careful to caution the carer not to '*throw them off a cliff top or try anything spectacular*'. Perhaps more helpful was the knowledge that a seabird's escape response when threatened is to dash into the sea, and so one must always approach an oiled bird from to seaward to cut off its escape.

Modern Advances

More advice on the complicated rituals of nursing a stricken seabird back to health, as understood in the 1960s, followed. Modern advice concerning how to deal with oiled birds is somewhat more nuanced and is readily available from the Royal Society for the Protection of Birds or the international Bird Rescue Research Centre. Nevertheless, the publication of this pamphlet so soon after the disaster is indicative of the impact of the *Torrey Canyon* wreck. The overriding impression one gets from reading contemporary accounts of the disaster is a feeling of helplessness at every level, from the government right down to the locals, who could only sit and watch as their beaches turned black.

Less than a year later the Liberian tanker *Ocean Eagle* ran aground off San Juan harbour, Puerto Rico, and broke in two, while the Greek tanker *General Colocotronis* struck a reef off Eleuthera Island in the Bahamas. Together, these wrecks stung the world into action, and never again would the response seem so futile, and the result so inevitable. Even so, in 1991, when the Iraqi army poured an unknown quantity of crude oil into the Kuwaiti Gulf, which produced a slick larger than any the world had seen before, perhaps 100 miles (160 km) long and 40 miles (64 km) wide, the situation seemed hopeless. Now, highly advanced booms can stop a slick in its tracks, carefully designed chemical solutions containing microbes that break down the oil quickly can be applied, and the oil itself can be skimmed from the surface by specially adapted ships, before being disposed of safely. Perhaps most importantly, tankers are now designed to retain their oil even under the extraordinary pressure of running aground.

We also now know that the immediate effects of spraying detergent on our beaches is not lasting, and that the algae, shellfish and other myriad examples of sea life that are killed or damaged by detergent are naturally replaced very quickly. That can only be said with the luxury of hindsight, however, and to understand the deep trauma suffered by the many thousands of people directly involved with the wreck of the *Torrey Canyon*, or directly affected by its spillage, we must remember that, at the time, many feared the oil and bleach would damage the coastal environment forever. We can easily empathize; it is exactly that unsettling feeling we are all increasingly familiar with, as we watch the hole in the ozone layer grow, the ice caps melt and the sea levels rise.

The *Kursk*

12 August 2000

Following the collapse of communism, the end of the Cold War and the economic liberalization of the former USSR, the Russian Navy was reduced to a shadow of its former self. Yet one tradition that had survived from the Soviet period – deep suspicion of the West and obsessive secrecy towards its own citizens – was to spell disaster for the 118 crewmen on board the nuclear cruise-missile submarine Kursk *when a military exercise in the Barents Sea in 2000 went terribly wrong.*

The Gulf Stream is a tongue of warm water that courses through the Atlantic Ocean. It originates in the Gulf of Mexico, exits into the Atlantic through the Florida Straits, and then stretches to northern Europe where it makes the waters warmer than they naturally should be. It even laps around the northern coast of Norway and reaches the north coast of Russia. This is a land of polar winters, where the few hours of daylight provide little warmth.

The regular and severe sub-zero temperatures cause regular snow storms, with great banks of snow and ice forming on land. But the sea never freezes; it is heated by the longest tentacles of the Gulf Stream. It was a discovery that catapulted Russia into becoming a naval power of global reach in the early years of the 20th century.

Until 1894 Russia's only naval bases were in the Baltic, the Black Sea and the Pacific. Both the Baltic and the Black Seas have very narrow, and easily defendable, gateways: access from the Baltic to the North Sea is guarded by the three Danish Straits (the Lille Bælt, Store Bælt and the Øresund), while access from the Black Sea to the Mediterranean is guarded by the Bosporus and then by the Dardanelles. The Pacific coast of Russia, moreover, is thousands of miles from the busy highways of the Atlantic. The Russian Navy, therefore, never had easy and guaranteed access to the Atlantic until it was found, at the very end of the 19th century, that her northern shoreline was equipped with large and deep natural harbours that never froze. By 1922 the Fleet of the Northern Seas was established and Russia was freed from her strategic imprisonment. Renamed the Northern Fleet by Stalin in the 1930s, by the 1980s its facilities on the Kola Peninsula in northern Russia contained the greatest concentration of naval power anywhere in the world, with as many as 180 submarines, operating from 12 bases on a coastline 300 miles (480 km) long.

A Routine Exercise Goes Awry

The Russian Northern Fleet conducts extensive military exercises every year, usually in August, and in 2000 that fleet prepared for the largest exercise in a decade. None of this was a surprise to the British or Americans, who have closely monitored Russian activity there since the 1950s; not only are those desolate waters populated by a Russian fleet of tremendous strength, but also by British and American submarines and numerous intelligence craft. Very little occurs there that is unobserved, and something as momentous as an explosion measuring 3.5 on the Richter scale – the equivalent of a minor earthquake – caused a great deal of concern in many countries. It was picked up by two American submarines, an American intelligence vessel, NATO surveillance planes, a Norwegian intelligence vessel and a Norwegian seismological station. The Russians also picked up an explosion, but they were so close that they also heard the subsequent sound of flooding from an area of sea where one of Russia's greatest submarines, the *Kursk*, was last known to be.

The *Kursk* was 5 storeys high, 154 metres (505 ft) long and 18.2 metres (59 ft) wide; a state-of-the-art *Oscar II*-class nuclear submarine (as designated by NATO) that cost over $1 billion to build. She could travel at over 30 knots (59 km/h) when submerged. She carried 118 sailors and was armed with 18 torpedoes and 23 cruise missiles. Part of her role in the exercises was to fire a dummy torpedo at a group of Russian ships, but the torpedo she was going to fire was over 25 years old. She was, moreover, under strict orders to report the results of the test as soon as they were known, but the Russian admiral, Vyacheslav Popov (b.1937), had received no news from the *Kursk* since the explosion had been heard. The Russians, therefore, had a shrewd idea that a disaster had befallen the *Kursk*; perhaps they even suspected that her elderly torpedo had exploded prematurely. It would not be long before the British, American and NATO surveillance forces worked it out for themselves.

At this stage, the disaster was already complex: 118 men, perhaps alive, perhaps dead, lay on the bottom of the Barents Sea, 58 fathoms (108 metres/350 ft) below the surface, with a nuclear reactor surrounded by torpedoes and cruise missiles. A rescue mission had to be launched to save any crew still alive, and a salvage operation instigated to raise the nuclear reactor. This might appear a daunting challenge, but it must be remembered that certain sections of the British and American navies are trained for exactly this sort of situation, and with military resources already in the vicinity it would not have taken long for a relief operation to be implemented. Now, however, the sinking of the *Kursk* took a significant turn that would entwine an already tragic tale with the vagaries of Russian political and military culture, and condemn her crew – some of whom were still clinging to life, submerged up to their waists in a watertight chamber – to a terrible death.

Fatal Inaction

Admiral Popov did not order an immediate search for the *Kursk*. In fact he took no action for 12 hours. In his eyes there was little point. There were no divers in the Russian navy trained to operate at that depth, nor were there any diving bells. The Russian Navy's only modern operational mini-subs capable of working at that depth were thousands of miles away in the North Atlantic. When finally the rescue mission was put in train it took 12 hours for ships with obsolete rescue submarines to even reach the site. There was no official announcement and family members of the *Kursk*'s crew came to hear of the disaster through snippets of gossip. The story soon spread to the Western press, but still it was clouded in confusion. The first assumption on all sides was that there had been a collision with an American submarine; it would not have been the first time. In fact the Russians claimed there had been 21 such collisions in the previous 30 years with US vessels alone. Collisions with British submarines were also not unheard of. Indeed it was this explanation, completely unsubstantiated, that the Russians officially chose to take to the press when they finally broke the story.

The British offered help in the form of their highly advanced rescue submarine and specially trained personnel, but were met with only a polite acknowledgement of their offer. So too were offers made – and turned down – from the United States, Norway, Canada, France, Germany, Israel, Italy, the Netherlands, Japan and Sweden. Meanwhile, images of the Russian rescue operation, struggling with rusty ships, were beamed across the world. And as the crisis unfolded, now with alarming international implications, the Russian president Vladimir Putin (b.1952) chose to stay by the beach at Sochi on the Black Sea, where he was enjoying a summer holiday. He made no comment for four days and the Russian media fumed. One editorial read:

> *'For the fifth day the whole of Russia, with its heart trembling, is keeping a fearful eye on the Barents Sea drama. From the very first moment Western leaders expressed their compassion and offered their help. Only our President stays silent. The Supreme Commander of our Armed Forces has nothing to say.'*

By now, however, the crew of the *Kursk* were all dead, but on the surface everyone believed that they could still be alive. Indeed, the official Russian line on Tuesday 15

August, four days after the explosion, was that their oxygen would run out at midnight on Friday the 18th. Russian claims that knocking could be heard from the hull, that diving bells had been lowered, and that mini-subs had explored the site were all untrue. They were, in truth, paralysed by indecision, and a lack of experience, knowledge and equipment. They were also loath to allow foreign submersibles anywhere near their stricken submarine and even more loath to admit their inability to save their own men. Even when foreign help was finally accepted, from Norway, crucial information such as the design of the aft escape hatch – which a rescue sub would have to link to – was difficult to obtain. The rivalries born from the Cold War, largely expressed through the fierce protection of military secrets against foreign intelligence, were still very much alive. Nevertheless, in an operation that lasted only six hours, Norwegian divers were able to confirm that every chamber aboard the submarine was flooded and that there were no survivors. By then the Russians had been prevaricating for seven days.

The Norwegian divers also confirmed that almost a quarter of the submarine's length at the bow, where the torpedoes were kept, had been devastated. Those who had heard the initial explosion began to consider the possibility that a faulty torpedo had exploded, which had then detonated a number of others. The Russians, who began to accept this theory publicly, used it to calm relatives of the crew by arguing that they would all have died quickly. It was an ill-judged strategy. A public meeting was organized, designed to bring the relatives of the *Kursk* tragedy in contact with the Russian leaders, but it only served to make matters worse. The world watched in disbelief as the images of that meeting were beamed to television sets worldwide. A disturbance at the rear of the hall suddenly became the focus of everyone's attention. It was the grieving mother of Senior Lieutenant Sergei Tylik, demanding answers and raging at the men who let her son die. In front of the cameras, she was restrained and injected with a sedative. Her body crumpled and she was dragged out: the meeting designed to repair the reputation of the Russian leadership in the eyes of their own public and the wider world only served to damage it further still. And it had not yet reached its lowest ebb.

The Truth Emerges

When the *Kursk* was raised in an operation costing over $65 million, the bodies of the crewmen were found, and the true story of the sinking began to fall into place. It was immediately clear that the whole crew had not died quickly, as the Russians claimed, but that 23 of them had survived in the chamber nearest the stern of the submarine for some time, all of them unharmed by the explosions. They had cowered in a compartment 8 metres long by

3 metres wide (26 ft x 10 ft); freezing, waiting for a rescue mission that would never come. Captain Third-Rank Dmitri Kolesnikov took charge, was given a scrap of paper, and made a few notes by the light of the luminous hands of his watch. He took a roll call and then wrote the following note to his wife:

> *'It's too dark here to write, but I'll try to feel. It looks like we have no chance, 10–20 percent. Will hope that someone will read this. Here is the list of those present from the other sections, who are now in the ninth and will try to escape. Hello to everybody, don't despair. Kolesnikov.'*

He carefully wrapped it in plastic and put it in his breast pocket above his heart. He was found with his hand shielding it. Other sailors also composed brief notes. Those that were written at the start of their incarceration in the tiny compartment are lucid, but as the oxygen ran out and hypothermia set in, the writing becomes increasingly incoherent. Senior Midshipman Andrei Borisov wrote: '*If you are reading this note it means I am dead. But your lives will carry on and I ask that my son becomes a true man, like I used to be.*'

It took less than a year for a clear picture of the events that led up to this terrible situation to emerge. An elderly torpedo had malfunctioned as components within it had deteriorated, leaking a particularly volatile fuel into the casing. It exploded as it was being loaded into its firing tube, creating a fireball that burned at 8000°C, hot enough to vaporize flesh. All of those sailors in the first two compartments died instantly. Those who survived all moved to the stern, away from the smoke. They were unable to release the emergency distress buoy, as its restraints had never been released since the submarine's initial construction eight years previously.

Less than three minutes after the first explosion, the second blast – the colossal explosion picked up by the many listening devices in the area – struck. Every sailor in the first five compartments died instantly. A flash fire raged in the first third of the hull, and sea water poured into the great rents in her hull, forcing her towards the sea bed at great speed. She rolled onto her side as she plummeted. Those responsible for the nuclear reactors sealed themselves into their reinforced section of hull to prevent any leakage and condemned themselves to death. The rest who were free to move scrambled into the rearmost compartment. They survived there until 7 or 8 p.m., a full 12 hours after the initial explosion.

They died as they attempted to renew the plates on the oxygen-regeneration unit, which are highly volatile and can explode into flame if they are brought into contact with even the tiniest drop of liquid. The sailors knew this, and it was a desperate risk that they had to take. If they did not renew the plates they would all shortly die of asphyxiation.

Perhaps they left it too late; perhaps their brains were slow from carbon dioxide poisoning, and they were unable to control their stiff and shaking arms. The plates caught fire. Three of the sailors created a protective barrier with their own bodies, shielding their crewmates from the chemical fire. They died from terrible chemical burns to their upper torsos, as they were submerged up to the waist. One man's gas mask melted onto his face. The remaining sailors did not die from that fire, but from asphyxiation as the flames used up all of the remaining oxygen. They would have slipped into a coma and died peacefully.

Systematic Failings

The sinking of the *Kursk* highlighted to the wider world much of what was wrong with post-Soviet Russia. One of the most significant facts about the wreck is that it is not an isolated instance of a Russian naval disaster. If one takes a broader view, the wreck of the *Kursk* was just one of a whole host of accidents to befall Russian submarines in the 20th century. These vessels were built too hastily or forced to operate with defective equipment to satisfy the Russian desire to compete with US and British naval power. The fate of the *Kursk*'s crew is mirrored throughout the history of the Soviet Navy, as time and again men have displayed levels of gallantry and self-sacrifice against odds stacked against them by their own government.

The Russian Navy has been systematically starved of funds. Only 40 submarines were active by the mid-1990s. Auxiliary and stand-by forces, such as submarine rescue, were cut back to keep those submarines at sea, and the submarines themselves, still examples of the most beautifully engineered craft in the world, were not maintained as they should have been. Few ever saw a dry dock and all began to rust. One's heart breaks for their sailors. In the words of a Russian Orthodox priest from the bleak naval base of Vidyayevo, home to the *Kursk*'s crew and families: '*There is no fleet in the world where there is such sacrifice. You leave shabby Vidyayevo and sail in a submarine to defend a country which exists more as an ideal than in reality.*'

The immediate government response to the wreck was unsurprising; indeed one need only look at Russia's mishandling of the 1986 Chernobyl disaster to find a clear antecedent. Then, it failed to prevent 8 tons of radioactive particles from polluting the Ukraine and neighbouring countries. The *Kursk* wreck was surrounded by wild claims exacerbated by disinformation from the government, channelled through the state-controlled press. One Russian newspaper neatly summed up the political impact of the *Kursk* disaster with the headline: '*The reputation of the Russian leadership lies on the bottom of the Barents Sea*'. Tellingly, such a comment could only have been made after 1990, when the Russian government first allowed the press some freedom. At the height of the Cold War, with limited press and investigative capability, the submarine would have sunk, the story would have been buried, and no one would have known anything about it. It can only be hoped, then, that the transparency of this Russian failure through modern investigation, research and media will ensure that high-level response to a future catastrophe will be more decisive, if only for political reasons.

To salvage some international standing, Putin sacked 14 top naval officers, and particularly targeted those who had claimed that the sinking was caused by a collision. Indeed, Vice-Admiral Vladimir Dobroskochenko, the only senior officer not to have mentioned a collision, was promoted to replace the disgraced and demoted Popov. Putin himself, who maintained that his poor actions were themselves caused by inaccurate information given to him by the navy, was re-elected with a massive 71 percent of the vote in March 2004. Those who remembered the obsessive secrecy that crippled the *Kursk* rescue operations were surprised a year later when, in the wake of the 11 September attacks, Putin agreed to the establishment of coalition military bases in Russian territory to assist with the invasion of Afghanistan.

An Unexploded Nuclear Time-bomb

The wreck of the *Kursk* also focused global attention on the isolated area of the Barents Sea where she and her sister nuclear submarines operate. Less than a year later a special scientific report from a conference on the danger of nuclear contamination in the region was published. It makes shocking reading. It is currently one of the most un-radioactive places in Europe, with the only significant radiation being washed ashore from Sellafield in Britain. But the potential for radiation is harrowing. The Northern Fleet has 142 nuclear submarines and three battlecruisers in or out of service. Between them there are over 300 nuclear reactors, to which the two from the *Kursk* can now be added. There have not, as yet, been any other accidents, but the threat is real.

These reactors, moreover, only represent the nuclear waste from the navy. The Kola nuclear power station, on the Kola Peninsula which reaches out into the Barents Sea, has four nuclear reactors. Two of those are now old and unstable. Significantly, power from the Kola power station also keeps many of the decommissioned nuclear reactors in a stable refrigerated environment until they are taken away for disposal. There are, in addition, only poor storage facilities in that region for spent nuclear fuel. Two storage ponds at the Murmansk naval facility had to be abandoned in 1982 because they leaked. Much spent nuclear fuel is now stored in elderly barges. All of this contamination is either on, under or near the coastline; a coastline that supports an important fishing fleet harvesting rich stocks of cod, herring, plaice, salmon and catfish.

If the sinking of the *Kursk* is to have any silver lining, then perhaps it is the growing awareness of the potential threat of nuclear pollution that is posed by that area of northern Russia. Elsewhere in Russia where pollution has affected communities, residents have developed blood disorders and leukaemia; babies have been born with physical and neurological defects; some have died of radiation sickness. If the global awareness and political will generated by the wreck can be maintained to deal with the potential pollution safely and soon, then some good can be salvaged from a terrible, and perhaps avoidable tragedy, and the 70 children who lost their fathers in the *Kursk* disaster can look forward to a safer future.

Index